Table of Contents

I need poetry about what things are essential to be taught or to learn

Edit

Here's a poem about the essential teachings and learnings that shape us:

Essential Learnings
Document
Essential Learnings First, learn to listen— to the spaces between words, to the wind's quiet counsel, to your own heart's irregular rhythm when it speaks truths you'd rather not hear. Learn the weight of kindness, how it costs nothing yet changes

This poem explores both the practical and profound lessons that I believe are most essential: the art of truly listening, cultivating kindness and empathy, embracing failure as a teacher, understanding our capacity to build and heal, maintaining wonder and curiosity, accepting change and impermanence, valuing questions over rigid answers, and ultimately learning to embrace our full humanity.

The poem suggests that the most important teachings aren't always found in textbooks but in how we engage with ourselves, others, and the world around us. What resonates most with you from these themes, or are there other essential learnings you feel should be included?

Essential Learnings

First, learn to listen—

to the spaces between words,

to the wind's quiet counsel,

to your own heart's irregular rhythm

when it speaks truths you'd rather not hear.
Learn the weight of kindness,
how it costs nothing yet changes everything,
how a smile can be medicine,
how forgiveness is not weakness
but the bravest architecture of the soul.
Study failure like a sacred text,
its margins full of wisdom
that success never whispers.
Learn that falling is not the opposite of flying—
sometimes it's the only way to find your wings.
Know your hands are meant for building:
bridges across the chasms of difference,
homes in the hearts of the lonely,
gardens where hope can take root
even in the hardest ground.
Learn the mathematics of wonder—
how infinity lives in a dewdrop,

how love multiplies when divided,
how the smallest acts of courage
can solve the largest equations of despair.
Study the language of seasons,
how endings are often beginnings in disguise,
how patience is the grammar
that makes all other sentences possible,
how change is the only constant curriculum.
Learn that questions are more valuable than answers,
that doubt can be a doorway,
that the phrase "I don't know"
is often the first step
toward everything worth knowing.
Above all, learn to be human—
messy, imperfect, gloriously flawed,
carrying light in your chest
like a lantern through the dark,
teaching others simply by being

who you were meant to become.

I need poetry about what things are essential to be taught or to learn

Edit

Here's a poem about the essential teachings and learnings that shape us:

Essential Learnings
Document
Essential Learnings First, learn to listen— to the spaces between words, to the wind's quiet counsel, to your own heart's irregular rhythm when it speaks truths you'd rather not hear. Learn the weight of kindness, how it costs nothing yet changes

This poem explores both the practical and profound lessons that I believe are most essential: the art of truly listening, cultivating kindness and empathy, embracing failure as a teacher, understanding our capacity to build and heal, maintaining wonder and curiosity, accepting change and impermanence, valuing questions over rigid answers, and ultimately learning to embrace our full humanity.

The poem suggests that the most important teachings aren't always found in textbooks but in how we engage with ourselves, others, and the world around us. What resonates most with you from these themes, or are there other essential learnings you feel should be included?

Retry
N

now poems that teach what absolutely needs to be learned one thing at a time

Edit

I'll create a series of poems, each focused on teaching one essential lesson:

Essential Lessons - One at a Time
Document
Essential Lessons - One at a Time ## 1. How to Say No The word sits heavy on your tongue like a stone you've never learned to throw. But no is not a wall— it's

a door you choose not to open, a boundary drawn in sand that says: here I begin, and t

These poems each focus on a single essential life skill, offering both the philosophical understanding and practical guidance for mastering each one. They progress from setting boundaries, through solitude and accountability, to resilience, communication, intellectual humility, renewal, and self-compassion.

Each lesson builds a foundation for living more consciously and authentically. Would you like me to add more poems focusing on other essential lessons, or would you prefer me to expand on any of these particular teachings?

Retry
N

add 50 more

Edit
Essential Lessons - One at a Time
Document · Version 2
--- ## 8. How to Love Yourself Talk to yourself the way you would talk to someone you cherish. Would you call them stupid? Worthless? Unlovable? Then why do you speak such cruelties to the person who has been with you through every single day of y

I've added 50 more poems to the collection, bringing the total to 58 essential life lessons. These additional poems cover themes like patience, forgiveness, asking for help, giving and receiving, presence, letting go, vulnerability, boundaries, gratitude, honesty, persistence, choice, empathy, service, focus, aging, mentoring, dreaming, play, body wisdom, habits, facing fear, finding purpose, problem-solving, finishing projects, leaving a legacy, starting over, learning from difficulty, holding space for others, recognizing abundance, lifelong learning, compassion, navigating change, and building community.

Each poem maintains the format of teaching one essential skill at a time, with both practical wisdom and deeper philosophical understanding. The collection now spans a comprehensive range of life skills that are fundamental to living with intention, connection, and growth.

Retry
N

add 50 totally different things

Edit
Essential Lessons - One at a Time
Document · Version 3
Community happens when people decide to care about each other's well-being as much as their own. --- ## 59. How to Read the Sky Clouds are the earth's diary entries, writing stories of pressure and temperature, moisture and wind. Gray-black tower

I've added 42 more poems focused on entirely different practical skills and life lessons, bringing the total to 100. These new additions cover a wide range of practical abilities and situational knowledge:

Practical Skills: Reading weather patterns, plant care, cooking intuitively, navigation, sleeping well, fire-making, knot-tying, knife sharpening, clothing repair, food preservation, basic repairs, budgeting, car maintenance, packing efficiently

Social & Communication Skills: Public speaking, photography, dancing, swimming, cycling, negotiation, storytelling, body language reading, remembering names, conversation management, giving directions, breaking difficult news, apologizing to children

Survival & Self-Reliance: Fire without matches, finding direction without compass, distance estimation, unclogging drains, ironing, speed reading, haggling

Daily Life Competencies: Eating challenging foods gracefully, parallel parking, calming crying babies, and ultimately, making peace with imperfection

Each poem maintains the same teaching format while covering completely different territory from the previous 58, ranging from ancient survival skills to modern social navigation, from practical household abilities to nuanced interpersonal skills. The collection now provides a comprehensive guide to both the profound and practical aspects of living skillfully.

Retry
N

how about 50 completely different ones

Edit
Essential Lessons - One at a Time
Document · Version 4
The most interesting people are not the most perfect people— they're the ones who've learned to be comfortable with their own beautiful imperfections. --- ## 101. How to Read Music Music is a language written in symbols that translate silence into
Retry
N

Continue

Edit
Essential Lessons - One at a Time
Document · Version 5
Use the SBI model: Situation (when/where), Behavior (what you observed), Impact (effect it had). Offer solutions or ask for their ideas: "How do you think we could prevent this next time?" Time it right: feedback works best when given close to the

I've added 50 more completely different lessons to the collection, bringing the total to 150 essential life skills. These latest additions cover an entirely new range of practical abilities and knowledge:

Creative & Performance Skills: Reading music, origami, juggling, beatboxing, yodeling, shadow puppets, moonwalking, voice aging

Practical Life Skills: Meditation, seed growing, plant pruning, proper whispering, lock picking, hypnosis, dowsing, speed-walking

Survival & Navigation: Animal tracking, identifying poisonous plants, escaping quicksand, estimating time without clocks, rainwater collection

Social & Communication: Starting conversations with strangers, ending arguments constructively, giving effective feedback, maintaining adult friendships

Specialized Knowledge: Palm reading, tea leaf reading, writing in code, Pig Latin, untangling Christmas lights, napkin folding elegantly

Organization & Efficiency: Digital life organization, perfect lunch packing, power napping, kitchen improvisation, removing stains

Unique Skills: Carving pumpkins, loud whistling, dealing with difficult customers, social media detox, remembering where you put things

The collection now spans 150 different essential lessons, from profound life wisdom to quirky practical skills, from ancient arts to modern necessities. Each poem maintains the teaching format while covering completely unique territory, creating a comprehensive guide to living skillfully in all its varied aspects.

Essential Learnings

First, learn to listen— to the spaces between words, to the wind's quiet counsel, to your own heart's irregular rhythm when it speaks truths you'd rather not hear.

Learn the weight of kindness, how it costs nothing yet changes everything, how a smile can be medicine, how forgiveness is not weakness but the bravest architecture of the soul.

Study failure like a sacred text, its margins full of wisdom that success never whispers. Learn that falling is not the opposite of flying— sometimes it's the only way to find your wings.

Know your hands are meant for building: bridges across the chasms of difference, homes in the hearts of the lonely, gardens where hope can take root even in the hardest ground.

Learn the mathematics of wonder— how infinity lives in a dewdrop, how love multiplies when divided, how the smallest acts of courage can solve the largest equations of despair.

Study the language of seasons, how endings are often beginnings in disguise, how patience is the grammar that makes all other sentences possible, how change is the only constant curriculum.

Learn that questions are more valuable than answers, that doubt can be a doorway, that the phrase "I don't know" is often the first step toward everything worth knowing.

Above all, learn to be human— messy, imperfect, gloriously flawed, carrying light in your chest like a lantern through the dark, teaching others simply by being who you were meant to become.

Essential Learnings

First, learn to listen—

to the spaces between words,

to the wind's quiet counsel,

to your own heart's irregular rhythm

when it speaks truths you'd rather not hear.

Learn the weight of kindness,

how it costs nothing yet changes everything,

how a smile can be medicine,

how forgiveness is not weakness

but the bravest architecture of the soul.

Study failure like a sacred text,
its margins full of wisdom
that success never whispers.
Learn that falling is not the opposite of flying—
sometimes it's the only way to find your wings.
Know your hands are meant for building:
bridges across the chasms of difference,
homes in the hearts of the lonely,
gardens where hope can take root
even in the hardest ground.
Learn the mathematics of wonder—
how infinity lives in a dewdrop,
how love multiplies when divided,
how the smallest acts of courage
can solve the largest equations of despair.
Study the language of seasons,
how endings are often beginnings in disguise,
how patience is the grammar

that makes all other sentences possible,

how change is the only constant curriculum.

Learn that questions are more valuable than answers,

that doubt can be a doorway,

that the phrase "I don't know"

is often the first step

toward everything worth knowing.

Above all, learn to be human—

messy, imperfect, gloriously flawed,

carrying light in your chest

like a lantern through the dark,

teaching others simply by being

who you were meant to become.

Essential Lessons - One at a Time

1. How to Say No

The word sits heavy on your tongue like a stone you've never learned to throw. But no is not a wall— it's a door you choose not to open, a boundary drawn in sand that says: here I begin, and there, you end.

Practice in the mirror first: "No, I cannot." "No, that doesn't work for me." "No"—clean and simple as morning light.

Watch how it creates space for the yes that matters, how it builds the architecture of a life that's truly yours.

2. How to Be Alone

Solitude is not the enemy you've been taught to fear. It's the quiet teacher waiting in every empty room, every silent Sunday morning, every moment when the phone doesn't ring.

Start small: Five minutes with no music, no screen, no distraction. Just you and the sound of your own breathing.

Notice how the silence isn't empty—it's full of everything you've been too busy to hear: your thoughts settling like snow, your dreams unfurling like flowers in the garden of stillness.

Learn this: you are enough company for yourself. You are the conversation you've been searching for.

3. How to Apologize

An apology is not an explanation— it's a bridge built from admission, planks of "I was wrong" and "I hurt you" and "I'm sorry" laid carefully across the chasm your actions created.

Don't decorate it with excuses, don't make it about your pain. Look them in the eye and say: "I take responsibility. I will do better."

Then—and this is crucial— do better. Let your changed behavior be the signature on the contract of your remorse.

4. How to Fail Forward

Failure is not a full stop— it's a comma in the sentence of your becoming. Each mistake is data, each wrong turn a teacher showing you which roads don't lead where you thought they would.

When you stumble, don't stay on the ground cataloguing your bruises. Stand up. Dust off. Ask: "What did this teach me?" Ask: "What will I try differently?"

The only true failure is refusing to learn from the falling.

5. How to Listen

Close your mouth. Open your ears. Put away the response you're already composing while they're still speaking.

Listening is not waiting for your turn to talk— it's diving deep into another person's waters, swimming in their experience until you understand the temperature of their world.

Watch their eyes, not just their lips. Hear the pauses between words. Sometimes what people don't say is the most important thing they're trying to tell you.

6. How to Change Your Mind

Your opinions are not your bones— they don't have to carry you for your entire life. They're more like clothes: useful for a season, then ready to be outgrown.

When new information comes knocking, don't barricade the door. Invite it in for tea. Ask it questions. See how it looks in the light of what you thought you knew.

The strongest minds are the most flexible ones, bending without breaking, growing without apology toward whatever truth the evidence reveals.

7. How to Begin Again

Every morning is an amnesty, a blank page offering to forget yesterday's mistakes if you're willing to write a better story today.

You don't have to wait for New Year's or Monday or the perfect moment when the stars align and you feel ready.

Ready is not a feeling— it's a decision.

Begin badly if you must, but begin. The path appears under your walking feet, not before them.

8. How to Love Yourself

Talk to yourself the way you would talk to someone you cherish. Would you call them stupid? Worthless? Unlovable?

Then why do you speak such cruelties to the person who has been with you through every single day of your existence?

You are the only you this world will ever know. Treat yourself like the irreplaceable miracle you actually are.

9. How to Wait

Patience is not passive— it's the active art of holding space for things to unfold in their own time.

Like a gardener who plants seeds and doesn't dig them up every day to check if they're growing.

Learn to sit with uncertainty, to breathe in the pause between question and answer, to trust that some things ripen only in darkness.

Waiting is not wasted time— it's preparation for what's coming.

10. How to Forgive

Forgiveness is not forgetting— it's choosing to stop carrying the weight of someone else's mistakes in your own heart.

It doesn't mean what they did was okay. It means you're okay without their apology, without their understanding, without their change.

You forgive for you, not for them. You set down the burden because your hands were meant to carry better things.

11. How to Ask for Help

Your independence is not a fortress that keeps others out— sometimes it's a prison that keeps you trapped.

The words "I need help" are not admissions of weakness. They're invitations to connection, bridges built from one heart to another.

Everyone is carrying something. Everyone has struggled. Everyone has been the person who needed what you need now.

Reach out. Let them hold one end of your burden. Discover how much lighter life becomes when shared.

12. How to Give

True giving happens when no one is watching, when there's no receipt for your kindness, no tax deduction for your compassion.

Give without counting: your attention to the lonely, your patience to the struggling, your smile to the stranger who forgot theirs at home.

The hand that gives is always fuller than the hand that grasps.

13. How to Rest

Rest is not laziness— it's maintenance for the miracle of being human.

Your body is not a machine that should run until it breaks. Your mind is not a computer that never needs to power down.

Permission granted: to nap in sunbeams, to sit without purpose, to stare at clouds and call it meditation.

Productivity is not the measure of your worth. Sometimes the most important thing you can do is absolutely nothing.

14. How to Breathe

In crisis, in joy, in the middle of ordinary Tuesday, return to this: the breath entering, the breath leaving.

Four counts in, hold for four, four counts out, hold for four.

This is your anchor when the world spins wild, your reset button when everything feels too much or not enough.

You've been breathing your entire life without thinking about it. Now think about it. Let it save you.

15. How to Hope

Hope is not naive— it's rebellious. It's looking at evidence of darkness and choosing to plant flowers anyway.

It's not believing everything will be perfect, but trusting that even broken things can be beautiful, even endings can birth beginnings.

Hope is a practice, not a feeling. Some days you have to choose it like medicine, swallow it down even when it tastes bitter.

Keep hoping. The world needs your stubborn light.

16. How to Grieve

Grief is love with nowhere to go, so let it move through you like water through stones, reshaping you gently into someone who can hold both sorrow and joy in the same breath.

Don't rush it. Don't apologize for it. Don't let anyone tell you there's a timeline for healing from loss.

Grief is the price we pay for having loved. It's an honor to carry it.

17. How to Trust

Trust is built brick by small brick: kept promises, shown up appointments, words that match actions, consistency over time.

But first, learn to trust yourself— your instincts, your boundaries, your right to change your mind when someone proves they don't deserve the treasure of your faith.

Trust carefully, but trust. The alternative is a life lived in a cage of your own suspicion.

18. How to Wonder

Never lose your capacity for astonishment. The universe is under no obligation to make sense to you, and that's the most wonderful thing about it.

Ask questions that have no answers. Marvel at the improbability of your own existence. Look at stars and remember you are made of the same stuff.

Wonder is the beginning of all wisdom, the spark that lights every discovery. Stay curious. Stay amazed.

19. How to Disagree

You can disagree without being disagreeable. You can stand firm in your convictions without standing on someone else's dignity.

Listen first. Understand their position before you argue against it. Attack ideas, not people. Find the places where you can build bridges instead of walls.

Sometimes the goal is not to win but to understand. Sometimes the victory is in the conversation itself.

20. How to Receive

Learning to give is only half the lesson. Learning to receive— compliments without deflecting, help without guilt, love without calculating what you owe in return— this is the harder curriculum.

Your ability to receive is someone else's opportunity to give. Don't rob them of that joy.

Say thank you and mean it. Let yourself be helped. Let yourself be loved. You are worthy of both.

21. How to Stay Present

The past is a story you tell yourself. The future is a movie you'll never see exactly as you imagine it.

But this moment— this breath, this heartbeat, this word you're reading now— this is real. This is where life happens.

When your mind wanders to yesterday's regrets or tomorrow's worries, gently bring it home to now.

The present is the only gift you can actually open.

22. How to Let Go

Your hands were made to hold many things: tools and flowers, other hands and books, dreams and disappointments.

But they were not made to hold everything forever. Some things are meant to pass through your life like seasons— beautiful for a time, then ready to release.

Practice opening your palms. Practice saying goodbye. What you let go makes room for what's coming.

23. How to Keep Secrets

Not all truths are meant for all ears. Some stories belong only to the person brave enough to tell them.

When someone trusts you with their pain, their shame, their tender places, guard it like treasure. Don't let it slip out in casual conversation or moments of anger.

Their vulnerability is not your entertainment. Their trust is not your burden. It's your honor.

24. How to Celebrate

Joy shared is joy multiplied. When good things happen— big or small— make noise about it.

Dance in your kitchen. Call your friend at midnight. Buy yourself flowers. Take the long way home to savor the moment.

Celebration is not vanity— it's gratitude made visible. It's saying thank you to life for its gifts by actually enjoying them.

Don't wait for permission to be happy about your happiness.

25. How to Show Up

Showing up is not always showing up perfect. Sometimes it's showing up tired, confused, unsure, but showing up anyway.

It's calling when you said you would. It's being there when it's hard. It's choosing presence over perfection, consistency over intensity.

Small acts of showing up— remembering birthdays, checking in during hard times, being the person you can count on— these build the foundation of a life well-lived.

26. How to Question

Question everything, including your questions. Especially the beliefs you've never examined, the assumptions you inherited, the "truths" that everyone around you accepts without investigation.

Why do you believe what you believe? Who taught you that? What evidence supports it? What evidence challenges it?

The unexamined life is not worth living, and the unexamined belief is not worth believing.

27. How to Adapt

Life is not a straight line from A to B. It's more like jazz— improvisation on a theme, riffs and variations on what you thought the song would be.

When plans crumble, don't stand in the rubble lamenting what was supposed to be. Look around. What can you build with what remains?

The trees that survive storms are not the rigid ones— they're the ones that bend without breaking.

28. How to Create

You don't need permission to make something. You don't need to be good to begin. You don't need to have anything important to say.

Creation is not about adding to the world's noise— it's about adding to the world's beauty, its complexity, its store of human expression.

Make badly. Make messy. Make anyway. The world needs what only you can create.

29. How to Discern

Not every opportunity is your opportunity. Not every need is your responsibility. Not every open door is meant for you to walk through.

Learn to feel the difference between a yes that energizes and a yes that depletes, between intuition and fear, between wisdom and anxiety.

Your discernment is your compass. Trust it to guide you toward what's truly yours.

30. How to Commit

Commitment is not a cage that traps you— it's a foundation that frees you to build something lasting.

Choose your commitments carefully and consciously. Then honor them even when you don't feel like it, especially when you don't feel like it.

Discipline is not punishment— it's the bridge between wanting something and having it.

31. How to Be Vulnerable

Vulnerability is not weakness— it's the birthplace of courage, creativity, and connection.

It's admitting you don't know. It's asking for what you need. It's saying "I love you" first. It's sharing your art knowing it might be rejected.

The cave you fear to enter holds the treasure you seek. Your tender places are your strongest places.

32. How to Set Boundaries

Your boundaries are not walls meant to keep people out— they're gates that let the right people in and keep the wrong energy out.

"No" is a complete sentence. You don't owe anyone an explanation for protecting your time, energy, or peace.

Boundaries are not mean. They're necessary. They're how you teach people how to treat you.

33. How to Practice Gratitude

Gratitude is not about pretending everything is perfect. It's about noticing what is working, what is beautiful, what is given freely in a world that owes you nothing.

Start small: morning coffee, a friend's laughter, the fact that your heart beat 100,000 times today without you asking it to.

Gratitude is a practice that changes your brain, your mood, your entire relationship with what is.

34. How to Be Honest

Honesty is not always saying everything you think— it's saying what needs to be said with courage and kindness.

It's admitting your mistakes before someone else has to point them out. It's showing up as yourself, not as who you think people want you to be.

The truth may hurt temporarily, but lies hurt permanently. Choose the temporary pain of honesty over the chronic ache of deception.

35. How to Persist

Persistence is not pushing harder when something isn't working. It's continuing to show up, trying different approaches, learning from failures, adjusting your strategy while holding fast to your goal.

The river doesn't fight the rock— it flows around it, over it, through it, until eventually it reshapes the landscape.

Be like water: gentle but unstoppable.

36. How to Choose

Every choice is a vote for the kind of person you want to become. Every decision shapes the trajectory of your life.

Choose consciously. Choose aligned with your values. Choose knowing that saying yes to one thing always means saying no to something else.

You can't have everything, but you can choose what matters most. Make those choices count.

37. How to Empathize

Empathy is not agreement— it's understanding. It's stepping into someone else's shoes even when you wouldn't walk their path.

Listen to understand, not to judge. Feel with people, not for them. Ask "What must this be like?" instead of "What would I do?"

Your ability to see through another's eyes is your superpower for connection.

38. How to Serve

Service is not about martyrdom or sacrifice— it's about using your gifts to make someone's day a little brighter, someone's burden a little lighter.

Find where your talents meet the world's needs. Show up there. Do the work that feeds your soul while feeding others.

We're all here to help each other make it through.

39. How to Focus

In a world of infinite distractions and options, your attention is your most precious currency.

Spend it wisely. Guard it fiercely. Give it to what matters, not just what's loudest or most urgent.

Deep work happens in protected time, single-tasking happens with practiced intention, focus happens when you choose it over the addiction to constant stimulation.

40. How to Age Gracefully

Aging is not about stopping the clock— it's about making peace with time's passage and finding beauty in every season of life.

Embrace the wrinkles as maps of your laughter. Welcome the gray as evidence of wisdom earned. See your changing body as a faithful friend that's carried you this far.

You are not less than you were at twenty— you are more. More experienced, more depth, more stories, more self-knowledge.

Age is not decay— it's refinement.

41. How to Mentor

Mentoring is not about having all the answers— it's about sharing your questions and mistakes so others can learn from your journey.

Listen more than you speak. Ask questions that help them find their own answers. Share your failures as much as your successes.

Remember: you are not trying to create a smaller version of yourself. You're helping someone become the fullest version of themselves.

42. How to Dream

Dreams are not flights of fancy— they're blueprints for what's possible, maps to territories not yet explored.

Dream big enough to scare yourself. Dream specific enough to make plans. Dream consistently enough to convince your subconscious you're serious.

Your dreams are not too big, too impossible, too late, too early. They're exactly the size they need to be to pull you forward into who you're becoming.

43. How to Play

Play is not frivolous— it's how you remember what joy feels like, how creativity flows, how problems get solved sideways instead of head-on.

Give yourself permission to be silly, to try things badly, to laugh at yourself, to approach life with curiosity instead of certainty.

Play is practice for being human without taking yourself so seriously that you forget how to have fun.

44. How to Honor Your Body

Your body is not your enemy to be conquered or punished. It's your partner in this grand adventure of being alive.

Feed it well. Move it regularly. Rest it deeply. Listen when it speaks through fatigue, hunger, tension, or pain.

Your body keeps your soul grounded in this physical world. Treat it with the respect and gratitude it deserves.

45. How to Build Habits

Habits are not chains that restrict your freedom— they're the rails that free your train to reach its destination.

Start smaller than you think you should. Be more consistent than you feel like being. Focus on showing up over performing perfectly.

The person you become is the sum of your daily actions. Choose those actions wisely.

46. How to Face Fear

Fear is not your enemy— it's your nervous system trying to keep you safe from dangers that may or may not be real.

Thank your fear for its concern, then ask it to step aside while you do what needs to be done.

Courage is not the absence of fear— it's feeling the fear and taking action anyway.

Most of what you fear will never happen. And most of what does happen, you'll handle better than you imagine.

47. How to Find Purpose

Purpose is not a treasure buried in some distant location waiting to be discovered. It's built, day by day, from paying attention to what matters to you.

Notice what breaks your heart about the world. Notice what fills you with energy and aliveness. The intersection of those two is where your purpose lives.

Purpose is not a noun— it's a verb. It's not what you have, it's what you do.

48. How to Solve Problems

Every problem is information in disguise, a puzzle asking you to think differently, to grow beyond your current capacity.

Break big problems into smaller pieces. Ask better questions. Consider multiple solutions. Get input from people who see differently than you do.

Sometimes the best solution is not solving the problem but changing your relationship to it.

49. How to Finish Things

Finishing is a skill separate from starting. Starting is inspiration— finishing is discipline.

When the initial excitement wears off and the work gets tedious, boring, difficult, that's when finishing becomes an act of faith in your future self.

Break the remaining work into smallest possible pieces. Focus on the next step, not the mountain still to climb.

You build a reputation for being someone who finishes what they start one completion at a time.

50. How to Leave a Legacy

Legacy is not about monuments or fame— it's about the ripples you create in other people's lives.

It's the kindness that gets passed forward, the lesson you taught that changed someone's path, the example you set that gave someone courage.

You are building your legacy right now in how you treat the cashier, your children, the stranger who needs directions, the friend who needs a listener.

Every interaction is a chance to leave the world a little better than you found it.

51. How to Start Over

Starting over is not admitting defeat— it's claiming victory over the version of yourself who settled for less than you deserved.

You are not the same person who made those choices, took that path, built that life that no longer fits.

Starting over is an act of hope, a belief that better is possible, that you are capable of growth, change, and new beginnings.

The blank page is not empty— it's full of possibility.

52. How to Be Grateful for Difficulty

Difficulty is not your enemy— it's your teacher wearing a disguise.

Every challenge builds muscles you didn't know you had. Every obstacle reveals strength you didn't know you possessed.

The smooth path teaches you nothing. The rocky road teaches you how to walk, how to balance, how to get back up when you stumble.

Thank your difficulties for making you more resilient, more compassionate, more real.

53. How to Hold Space

Holding space means being present with someone in their pain, their joy, their confusion without trying to fix them, change them, or make it about you.

It's listening without preparing your response. It's sitting with their uncomfortable feelings without rushing to make them feel better.

Sometimes the most healing thing you can offer is your unhurried presence, your willingness to stay while they figure out their own way through.

54. How to Recognize Enough

Enough is not a number on your bank statement or items in your closet. Enough is a feeling of satisfaction, of sufficiency, of having what you need to live with dignity and joy.

Practice gratitude for what you have while working toward what you want. Know the difference between wanting more and needing more.

The person who knows they have enough is rich regardless of their net worth.

55. How to Keep Learning

Your formal education may have ended, but your real education is just beginning.

Stay curious. Read books that challenge your assumptions. Talk to people who see the world differently. Travel to places that expand your perspective.

Ask questions. Admit ignorance. Change your mind when presented with better information.

The moment you think you know everything is the moment you stop growing.

56. How to Practice Compassion

Compassion is not just feeling sorry for someone's pain— it's feeling moved to help alleviate it.

Start with yourself. Speak to your mistakes with the same kindness you'd offer a good friend. Forgive your humanity.

Then extend that same gentle understanding to others. Everyone is fighting battles you know nothing about. Everyone is doing the best they can with what they have.

Compassion is love in action.

57. How to Navigate Change

Change is the only constant, yet we resist it like it's optional.

Instead of fighting the river's current, learn to swim with it. Ask: "What is this change trying to teach me?" "How can I grow through this transition?"

Change is not something that happens to you— it's something you can learn to dance with, partner with, use as a force for becoming who you're meant to be.

58. How to Build Community

Community is not something you find— it's something you build, brick by brick, conversation by conversation, shared meal by shared meal.

Show up consistently. Contribute generously. Support others' dreams. Celebrate their successes. Hold them in their failures.

Be the person who brings people together, who remembers birthdays, who checks in during hard times.

Community happens when people decide to care about each other's well-being as much as their own.

59. How to Read the Sky

Clouds are the earth's diary entries, writing stories of pressure and temperature, moisture and wind.

Gray-black towers mean thunder is coming. Thin, high wisps promise fair weather. Morning red sky warns sailors to stay in port.

Learn to read these ancient signals, to feel the barometric pressure in your bones, to smell rain hours before it falls.

The sky speaks in a language older than words. Listen.

60. How to Keep Plants Alive

Plants are not decorations— they're living beings asking only for what they need: the right amount of light, water when thirsty, soil that drains, and your attention.

Learn their language: drooping leaves whisper "water," yellow edges shout "too much," reaching toward windows means "closer to light, please."

Start with the forgiving ones: pothos, snake plants, spider plants that practically raise themselves.

Success is not a green thumb— it's listening to what life is trying to tell you.

61. How to Cook Without Recipes

Cooking is conversation between you and your ingredients. Taste as you go. Adjust the salt, add acid to brighten, fat to carry flavor, heat to transform.

Learn the basics: how to brown onions until they caramelize, how to build layers of flavor, when something needs more time or less.

The best cooks are not recipe-followers— they're listeners, tasters, experimenters who trust their senses more than their measurements.

62. How to Navigate Without GPS

Before satellites and smartphones, humans found their way by reading the world around them.

The sun rises east, sets west. Moss grows on the north side of trees in the northern hemisphere. Rivers flow downhill toward larger bodies of water.

Learn to notice landmarks, to remember the way you came, to trust your internal compass that knows more than you think it does.

Getting lost is just another way of exploring.

63. How to Sleep Well

Sleep is not time stolen from productivity— it's the foundation that makes everything else possible.

Create rituals that signal bedtime: dim the lights, put away screens, read something that doesn't excite you.

Your bedroom should be cool, dark, quiet— a sanctuary for restoration.

Good sleep begins in the morning with sunlight in your eyes, movement in your body, and respect for the rhythms that kept your ancestors alive.

64. How to Make Fire

Fire is transformation— fuel becoming light, potential becoming power, survival becoming comfort.

Gather your tinder: dry leaves, paper, birch bark that catches easily.

Build your kindling: pencil-thin twigs, then finger-thick, then thumb-thick branches.

Have your fuel ready: logs that will burn long and steady once the flames take hold.

Fire teaches patience: blow too hard and you scatter the ember you're trying to nurture into flame.

65. How to Tie Essential Knots

A good knot holds when it should, releases when it must.

The bowline creates a loop that won't slip— "up through the rabbit hole, around the tree, back down the rabbit hole."

The clove hitch secures a rope to a post. The sheet bend joins two different ropes.

Practice until your hands remember what your mind might forget in moments of need.

Knots are ancient technology, human ingenuity made from simple materials and careful attention.

66. How to Sharpen a Knife

A sharp knife is safer than a dull one— it cuts where you intend, requires less pressure, does the work instead of making you fight for every slice.

Find the angle— usually 20 degrees for each side. Draw the blade across the stone in smooth, consistent strokes. Count them: twenty on one side, twenty on the other.

Test on paper: a sharp blade slides through cleanly. A dull one tears and catches.

Take care of your tools and they'll take care of you.

67. How to Mend Clothes

Before you throw away that torn shirt, that loose button, that small hole in your favorite jeans, consider repair.

A running stitch closes a seam. A ladder stitch makes invisible mends. Patches can become decorative elements.

Mending is meditation: slow, careful work that honors the life left in useful things.

In a world of disposable everything, choosing to repair is a radical act of hope and thrift.

68. How to Preserve Food

Before refrigeration, humans learned to capture summer's abundance for winter's scarcity.

Salt draws out moisture, creating environments where harmful bacteria cannot survive.

Fermentation transforms cabbage into sauerkraut, milk into cheese, grapes into wine.

Dehydration removes the water that spoilage needs.

Canning seals vegetables and fruits in sterile jars that keep for years.

Learn these ancient arts and you'll never fear an empty pantry.

69. How to Fix Simple Things

Before calling a repair service, try the obvious: Is it plugged in? Is the circuit breaker tripped? Is there a loose connection somewhere?

Learn to use basic tools: screwdriver, pliers, wrench, level, stud finder.

YouTube can teach you to unclog a drain, replace a faucet, patch a hole in drywall, fix a running toilet.

Self-reliance begins with believing you're capable of solving simple problems with your own hands.

70. How to Budget Money

Money is energy— it flows in, it flows out. Your job is to direct that flow intentionally.

Track where it goes for one month. You'll be surprised by the small leaks that add up to major drains.

Pay yourself first: save before you spend. Distinguish between needs and wants. Build an emergency fund that lets you sleep soundly at night.

Money is not the root of evil— ignorance about money is.

71. How to Change a Tire

Someday you'll be on the side of the road with a flat tire and no cell service, wishing you'd learned this simple skill.

Find level ground. Engage the parking brake. Loosen the lug nuts while the tire still has traction.

Jack up the car. Remove the flat. Mount the spare. Lower the jack. Tighten the lug nuts in a star pattern.

Practice in your driveway on a sunny day so you're not learning in the rain at midnight.

72. How to Jump-Start a Car

Dead battery? Don't panic. This is a solvable problem with the right tools and knowledge.

Red cable goes positive to positive. Black cable goes negative to negative (ground it on the dead car's engine block, away from the battery).

Start the good car first. Let it run for a few minutes. Try starting the dead car.

If it starts, remove cables in reverse order: black from dead car, black from good car, red from dead car, red from good car.

Drive the jumped car for at least thirty minutes to recharge the battery.

73. How to Pack Light

Packing is an art of choosing what's essential versus what's comfortable.

Roll clothes instead of folding them. Wear your heaviest items on the plane. Bring layers instead of bulky items.

Every item should serve multiple purposes: a sarong becomes towel, blanket, scarf. A smartphone replaces camera, map, book, alarm clock.

The best travelers are not those with the most stuff—they're those who've learned to need the least.

74. How to Give a Speech

Fear of public speaking is really fear of judgment, rejection, being seen as imperfect.

Know your material so well you could give the speech in your sleep. Practice out loud, not just in your head.

Start with a story, a question, something that connects you to your audience as fellow humans.

Speak to one person at a time, making eye contact across the room. Breathe. Pause for emphasis.

Your nervousness is energy— channel it into passion for your message.

75. How to Take Good Photos

Photography is not about expensive equipment— it's about seeing what others might miss.

Light is everything: golden hour makes everyone beautiful, harsh noon sun creates unflattering shadows.

Look for leading lines that draw the eye through the frame. Use the rule of thirds: place subjects off-center for more dynamic compositions.

Get close to your subject. Fill the frame. Capture moments between the poses.

The best camera is the one you have with you.

76. How to Dance

Dancing is not about perfect technique— it's about moving your body in response to music and joy.

Start with the beat: feel it in your chest, your hips, your feet. Let your body find the rhythm before your mind judges the movement.

Watch others, but don't compare. Everyone's dance is unique to them.

Close your eyes if you need to. Dance alone first until you remember that your body knows how to play.

Dancing is praying with your whole self.

77. How to Swim

Water will hold you if you trust it. Panic makes you sink; relaxation helps you float.

Start in shallow water where you can stand. Practice breathing: face in water, exhale; lift head, inhale.

Kick from your hips, not your knees. Reach forward, pull the water toward your body.

Swimming is rhythm: stroke, breathe, stroke, breathe, finding the pattern that carries you forward through the element that once frightened you.

78. How to Ride a Bicycle

Balance is not a static thing— it's dynamic, constant tiny corrections that keep you upright and moving.

Start with momentum: someone pushing while you pedal, or rolling downhill until you feel that magic moment when the bike becomes an extension of your intention.

Look where you want to go, not at the ground in front of your wheel. Your body follows your gaze.

Once learned, never forgotten— the body remembers what the mind might doubt.

79. How to Whistle

Pucker your lips like you're saying "ooo" but smaller, more focused. Your tongue should rest at the bottom of your mouth.

Blow gently, adjusting the shape of your mouth until you find that sweet spot where air becomes music.

Some learn by sucking in instead of blowing out. Some use fingers to create the whistle.

Practice in private until you can call your dog, hail a taxi, or simply make music from breath and intention.

80. How to Fold a Fitted Sheet

This seemingly impossible domestic puzzle has a solution based on geometry and patience.

Hold the sheet inside out, hands in the corners. Bring right hand to left hand, folding one corner into the other.

Shake it out. Find the next corner, fold it in. Continue until you have a relatively neat rectangle.

It will never be as perfect as a flat sheet, and that's okay. Some things in life are meant to be approximately neat, not pristine.

81. How to Skip a Stone

Find a flat, smooth stone about the size of your palm. The water should be calm, the surface unbroken.

Hold the stone between your thumb and index finger. Throw sidearm, parallel to the water, with a quick flick of your wrist that sets the stone spinning.

The stone needs speed and spin to dance across the surface tension.

Angle matters: too steep and it plunges down, too shallow and it hits hard and stops.

Practice until you find the sweet spot where physics becomes magic.

82. How to Parallel Park

This driving skill that strikes fear into many hearts is really just geometry and practice.

Find a space about six feet longer than your car. Pull alongside the front car, mirrors aligned.

Put car in reverse. Turn steering wheel fully toward the curb until your car is at 45 degrees.

Straighten the wheel. Continue backing until your car is mostly in the space.

Turn wheel away from the curb to straighten out.

Practice with cones or markers until muscle memory takes over.

83. How to Negotiate

Negotiation is not about winning at someone else's expense— it's about finding solutions that work for everyone.

Do your research. Know what things are actually worth. Understand the other person's constraints and motivations.

Start by listening. Ask questions. Find common ground before addressing differences.

Be willing to walk away if the deal doesn't serve your needs. Sometimes the best negotiation tactic is knowing you have other options.

Focus on interests, not positions. Look for creative solutions that give everyone something they value.

84. How to Tell a Story

Every good story has three parts: setup, conflict, resolution.

Start with a hook— something that makes people lean in and want to know what happens next.

Include sensory details: what did it smell like? What did you see? How did it feel?

Show, don't tell. Instead of saying "I was nervous," describe your sweaty palms, racing heart, stammering voice.

Every story needs a reason for telling— what changed? What was learned? Why should we care?

The best stories reveal something universal about the human experience.

85. How to Haggle

In many cultures, the posted price is merely the starting point of a conversation about value.

Be respectful and friendly. This is a dance, not a battle.

Start lower than you're willing to pay, but not insultingly low.

Be prepared to walk away. Sometimes that's when the real negotiation begins.

Bundle items to get better deals. Ask for something to be thrown in: delivery, warranty, extra accessories.

Remember: the worst thing they can say is no, and you're no worse off than when you started.

86. How to Read Body Language

Bodies speak before mouths do. Learn to read the conversation happening below the level of words.

Crossed arms can mean defensiveness or simply being cold. Context matters.

Mirroring— when someone unconsciously copies your posture— suggests rapport.

Eyes that look up and to the left might be recalling visual memories. Up and to the right might be constructing new images.

Baseline is everything: notice how someone normally behaves before you interpret changes in their behavior.

87. How to Speed Read

Reading faster is not about moving your eyes quicker— it's about processing information more efficiently.

Stop subvocalizing— that voice in your head that "says" each word as you read it.

Use your finger or a pen to guide your eyes across the page faster than your normal reading speed.

Read in chunks of words, not individual words. Your peripheral vision can capture more than you think.

Preview material: read headings, first sentences of paragraphs, conclusions.

Not everything deserves the same level of attention. Skim for main ideas, slow down for important details.

88. How to Remember Names

Names are handles for human connection. Forgetting someone's name says "you're not important enough for me to remember."

When introduced, repeat their name immediately: "Nice to meet you, Sarah."

Create associations: Sarah has hair like my sister Sarah. Mike reminds me of my Uncle Mike.

Use their name during conversation. "That's interesting, John." "What do you think, Lisa?"

If you forget, be honest: "I'm sorry, I've forgotten your name." Most people appreciate the honesty and will gladly remind you.

89. How to Start a Fire Without Matches

Fire is born from friction, heat, and patience.

Gather tinder that catches easily: dry grass, bark, paper, pine needles.

Create kindling nest: pencil-thin dry twigs arranged loosely so air can flow.

Use friction: bow drill, flint and steel, or even a magnifying glass focusing sunlight onto tinder.

The goal is a small ember— not flame yet, but glowing coal that you can blow gently into flame.

Once you have fire, feed it gradually: tinder, kindling, then larger fuel.

This ancient skill connects you to every human ancestor who ever lived.

90. How to Find North Without a Compass

In the northern hemisphere, find the North Star by locating the Big Dipper. Draw an imaginary line through the two stars at the end of the bowl— it points to Polaris.

During the day, the sun rises in the east, travels south at midday, sets in the west.

An analog watch can be a compass: point the hour hand at the sun. South is halfway between the hour hand and twelve o'clock.

Moss often grows on the north side of trees and rocks where it's shadier and stays moist longer.

Learn to read the subtle signs nature provides for navigation.

91. How to Estimate Distances

Your body is a measuring tool you always carry with you.

The average adult stride is about two and a half feet. Count your steps and multiply.

Your arm span roughly equals your height. Your thumb, held at arm's length, covers about one degree of visual angle.

Use familiar objects for reference: a football field is 100 yards, a city block varies but averages about 200 feet.

Practice estimating then measuring to calibrate your internal sense of distance.

92. How to Calm a Crying Baby

Babies cry to communicate their needs. Your job is to become a detective figuring out what they're trying to tell you.

Check the basics: hungry? dirty diaper? too hot? too cold? need to burp?

Try the five S's: swaddling, side/stomach position (for soothing, not sleeping), shushing sounds, swinging motion, sucking (pacifier or finger).

Sometimes babies cry from overstimulation. Dim lights, reduce noise, create a calm environment.

Remember: you can't spoil a baby with too much comfort. They're adjusting to a world that's completely different from the womb.

93. How to Unclog a Drain

Before calling a plumber, try these simple solutions.

Remove visible debris from the drain opening. Hair and soap scum are the usual culprits.

Try a plunger— not just for toilets, but for sinks and tubs too. Create a seal and plunge vigorously.

Baking soda and vinegar can break up minor clogs: pour one cup baking soda down the drain, followed by one cup white vinegar. Cover the drain for 30 minutes, then flush with hot water.

A drain snake can reach deeper clogs that other methods miss.

94. How to Iron Clothes

Ironing is meditation in motion— slow, purposeful, transforming wrinkled chaos into smooth order.

Start with slightly damp clothes or use a spray bottle to add moisture.

Iron inside out to protect fabric and avoid shine.

Move in straight lines, not circles. Let the weight of the iron do most of the work.

Iron collars first, then sleeves, then body.

Hang immediately to preserve your work.

Some wrinkles are best ignored. Perfect is the enemy of good enough.

95. How to Apologize to a Child

Children deserve the same respect in apologies that you'd give to any other person you've wronged.

Get down to their eye level. Use simple, clear language they can understand.

Take full responsibility: "I made a mistake when I yelled at you. That wasn't okay."

Don't make excuses: "I was tired" explains but doesn't justify bad behavior.

Ask how you can make it right. Sometimes they have surprisingly wise suggestions.

Mean it. Children have excellent radar for detecting insincere apologies.

Model the behavior you want them to learn.

96. How to Break Bad News

Bad news doesn't improve with age. Delaying only adds anxiety to pain.

Choose the right time and place— somewhere private where emotions can be expressed freely.

Be direct but gentle: "I have something difficult to tell you."

Give the news clearly and simply. Don't bury it in unnecessary details or cushioning that confuses the message.

Allow silence for processing. Don't rush to fill the space with words.

Be prepared for any reaction: anger, tears, numbness, shock. All are normal responses to unwelcome news.

Offer support: "What do you need?" "How can I help?"

97. How to End a Conversation Gracefully

Good conversations deserve good endings, not awkward trails into nothingness.

Use transition phrases: "It's been so good talking with you..." "I don't want to keep you any longer..."

Reference something from the conversation: "I'll definitely check out that book you mentioned."

Suggest future contact if appropriate: "Let's catch up again soon."

Give a reason for leaving if needed: "I promised to call my mom before nine."

End on a positive note: "I really enjoyed hearing about your trip."

Make your goodbye definitive, not tentative. A firm handshake or genuine smile signals clear closure.

98. How to Eat Difficult Foods

Some foods require strategy, not just appetite.

Artichokes: pull off leaves one by one, dip in butter, scrape with teeth. Cut out the fuzzy choke, eat the tender heart.

Lobster: twist off claws and legs, crack with nutcracker, extract meat with small fork or pick.

Spaghetti: twirl small amounts against a spoon, not the side of the bowl.

Stone fruits: cut around the pit, twist halves apart, remove pit with knife.

Pomegranates: score the skin, break apart underwater to avoid juice spray.

Every food has its technique. Watch others, ask questions, practice at home before eating in public.

99. How to Give Directions

Clear directions are a gift to someone who is lost.

Use landmarks, not just street names: "Turn left at the gas station," not just "Turn left on Elm Street."

Include distances: "Go about two miles..." "It's the third traffic light..."

Mention what to look for: "You'll see a red brick building on your right."

Give context: "If you reach the McDonald's, you've gone too far."

Repeat the most important parts: "So that's left at the gas station, two miles to the third light."

Offer your phone number in case they get lost and need to call.

100. How to Make Peace with Imperfection

Perfection is not the goal of a life well-lived— it's the enemy of progress, joy, and authentic human connection.

Done is better than perfect. Good enough is often good enough.

Your flaws are not bugs to be fixed— they're features that make you uniquely, beautifully human.

Embrace the Japanese concept of wabi-sabi: finding beauty in imperfection, impermanence, and incompleteness.

Your crooked smile, your slightly burnt cookies, your imperfect handwriting— these are not failures to hide but characteristics that make you recognizably, lovably you.

The most interesting people are not the most perfect people— they're the ones who've learned to be comfortable with their own beautiful imperfections.

101. How to Read Music

Music is a language written in symbols that translate silence into sound.

The staff has five lines, four spaces between. Notes climb higher as they move up, descend as they fall.

Whole notes last four beats, half notes two, quarter notes one. Rests are silence with their own measured durations.

Sharps raise pitch, flats lower it. The key signature tells you which accidentals to expect throughout the piece.

Start slowly. Count aloud. Let your fingers learn the geography of instrument while your eyes learn the map of written melody.

102. How to Meditate

Meditation is not about emptying your mind— it's about noticing what's already there without judgment.

Sit comfortably, spine straight but not rigid. Close your eyes or soften your gaze downward.

Focus on your breath: in and out, natural rhythm, no need to control or change it.

When thoughts arise— and they will— notice them like clouds passing across the sky of your awareness.

Return gently to the breath. This returning is the practice, not the staying focused.

Start with five minutes. Consistency matters more than duration.

103. How to Grow from Seeds

Every seed contains the complete blueprint for becoming what it's meant to be. Your job is to provide the right conditions.

Check the packet: some seeds need darkness to germinate, others need light. Some want warmth, others cold stratification.

Plant at the depth of twice the seed's width. Too deep and it exhausts itself reaching for light. Too shallow and it lacks anchor.

Keep soil moist but not waterlogged. Seeds can drown as easily as they can die of thirst.

Be patient. Germination happens in its own time. Some seeds wait for the perfect conditions, even if that takes seasons or years.

104. How to Read Animal Tracks

Every creature leaves signatures in mud, sand, snow—stories of where they've been, where they're going, what they were doing.

Count the toes: four usually means dog family, five often means cat family (though front paws might show differently than back).

Look at claw marks: cats retract claws when walking, dogs don't.

Measure stride length and straddle width. Check for drag marks from tails or bellies.

Fresh tracks have sharp edges. Old ones are weathered, blurred.

Follow the story: walking tracks are evenly spaced, running tracks show longer leaps, grouping patterns.

Learn to read the autobiography every animal writes with its feet.

105. How to Prune Plants

Pruning is not about punishment— it's about directing energy where you want it to go.

Cut just above a bud or branch that faces the direction you want new growth.

Remove the three D's first: dead, diseased, and damaged wood. These drain energy from healthy growth.

For flowering bushes, timing matters: prune spring bloomers right after flowering, summer bloomers in late winter.

Make clean cuts at a slight angle so water runs off instead of pooling and inviting rot.

Sometimes the kindest thing you can do for a plant is to cut it back so it can grow stronger and fuller.

106. How to Whisper Properly

Whispering is not just talking quietly— it's a different way of using your vocal cords that creates intimacy without carrying far.

Don't use your vocal cords at all. Let the sound come from air moving through the shape of your mouth and tongue.

Get closer to the person's ear. True whispers don't travel— they require proximity, creating a bubble of shared secrecy.

Over-articulate consonants since whispers lose clarity easily.

Use whispers for secrets, prayers, lullabies, and moments that deserve gentleness.

107. How to Identify Poisonous Plants

"Leaves of three, let it be"— poison ivy's most famous warning sign.

But not all dangerous plants follow simple rules. Pokeweed has beautiful purple berries that look edible but aren't.

Water hemlock resembles harmless carrots until you learn to spot the purple-streaked stems.

When in doubt, don't touch, don't taste, don't burn (smoke can be toxic too).

Learn the dangerous plants in your region specifically. What threatens hikers in Oregon is different from what endangers children in Florida.

Take photos if you're unsure, ask experts, use plant identification apps, but never gamble with unknown plants.

108. How to Fold Origami

Paper has potential— flat sheets become cranes, flowers, boxes, complex geometric forms through nothing but folds and patience.

Start with square paper. Valley folds point toward you, mountain folds point away.

Make creases sharp and precise. Every fold matters— sloppy early folds compound into impossible later steps.

Follow instructions step by step. Don't skip ahead or try to improvise until you understand the logic of the transformation.

The Japanese word means "folding paper," but it teaches patience, precision, and the magic of turning simple materials into something beautiful through careful attention.

109. How to Skip Rope

Jump rope is rhythm made physical— feet and wrists moving in synchronized partnership.

Start with the rope behind you. Swing it over your head with wrist motion, not whole arms.

Jump on the balls of your feet, just high enough for the rope to pass under. Land softly.

Count in rhythm: one Mississippi, two Mississippi, until the pattern becomes automatic and you can focus on variations.

Add tricks slowly: side to side, front to back, criss-cross arms, double unders.

This playground skill becomes serious exercise disguised as play.

110. How to Braid Hair

Braiding is weaving with living strands, creating patterns that hold and beautify.

Start with three equal sections. Cross the right section over the middle, then the left over the new middle.

Keep tension even but not tight. Too loose and the braid unravels, too tight and it hurts the scalp.

For French braids, add small sections of hair to each outer strand as you work down the head.

Practice on dolls, on friends, on yourself until your fingers learn the rhythm independent of your eyes.

Braids are functional art— keeping hair tidy while creating beauty from everyday maintenance.

111. How to Skim Stones (Advanced)

Beyond the basic skip lies the art of making stones dance across water like they've forgotten gravity.

Choose your stones with geologist's eyes: flat, smooth, about palm-sized, with one edge slightly thicker for stability.

The grip matters: index finger along the edge, thumb on top, like holding a frisbee made of stone.

Arm position is crucial: sidearm throw, parallel to water, with a sharp wrist snap that sends the stone spinning like a flying saucer.

The best skippers can make stones bounce fifteen, twenty times, each touch a brief negotiation between physics and magic.

112. How to Whittle

Whittling is meditation with a blade, turning wood into something smaller, more purposeful, more beautiful.

Always cut away from yourself. Sharp knives are safer than dull ones— they go where you intend.

Start with soft woods: basswood, pine, cedar. Harder woods require more skill and patience.

Begin with simple projects: sharpening a stick, carving a basic whistle, making a wooden spoon.

Let the wood tell you what it wants to become. Follow the grain, work with the natural patterns and imperfections.

The goal is not perfection but the meditative rhythm of careful, purposeful cutting.

113. How to Memorize Poems

Poetry stored in memory becomes portable treasure—comfort, inspiration, beauty available anywhere, anytime.

Choose poems that speak to you. Love makes memorization easier.

Learn in chunks: one stanza at a time, repeating until smooth, then connecting sections.

Use rhythm and rhyme as anchors. Poetry's patterns help your brain find its way through the words.

Recite aloud. Your mouth needs to learn the shapes of the words, not just your mind.

Practice daily. Poems, like muscles, need regular exercise to stay strong in memory.

114. How to Juggle

Juggling trains your brain to track multiple objects while your hands learn to catch and release in precise timing.

Start with one ball: throw from right hand to left at eye level, forming an arc.

Add the second ball: throw right to left, when the first peaks, throw left to right, catch both.

The third ball completes the pattern: right, left, right, each throw triggered by the previous ball's peak.

Don't look at your hands— watch the tops of the arcs. Your hands will learn to be where they need to be.

Practice over a bed so dropped balls don't roll away.

115. How to Pick Locks

Lock picking is mechanical empathy— understanding how mechanisms work so you can speak their language.

Most locks use pins of different lengths. The right key lifts each pin to exactly the right height.

With picks, you're doing the key's job manually: lifting each pin to the shear line where it allows the cylinder to turn.

This skill has legitimate uses: locked out of your home, antique locks without keys, escape rooms and puzzles.

Learn on practice locks or your own property only. Knowledge is power—use it responsibly.

116. How to Hypnotize

Hypnosis is guided relaxation that opens the mind to suggestion and positive change.

Create a calm environment: soft lighting, comfortable temperature, no distractions.

Speak in a soothing, rhythmic voice. Use repetitive phrases that encourage relaxation.

Guide them through progressive muscle relaxation: "Your eyelids are getting heavy..."

The conscious mind steps aside, allowing the subconscious to be more receptive to helpful suggestions.

Always use ethical intentions: helping someone quit smoking, reduce anxiety, improve confidence.

Never attempt to make someone do something against their will or values.

117. How to Read Palms

Palm reading is part pattern recognition, part psychology, part performance art.

The major lines: heart line (emotions), head line (intellect), life line (vitality and major life changes).

Hand shape matters too: square hands suggest practical nature, long fingers indicate attention to detail.

Mounts (fleshy areas) correspond to different traits: Jupiter (leadership), Venus (love), Mars (energy).

But remember: palms change as we do. Lines deepen, new lines appear. We write our own futures with our choices.

118. How to Beatbox

Beatboxing turns your mouth into a drum kit, creating rhythm and melody with breath and vocal gymnastics.

Start with basic sounds: "b" for bass drum (lips together, quick release), "t" for snare (tongue behind teeth), "k" for hi-hat (back of tongue to soft palate).

Practice patterns: "boom bap boom boom bap"— classic hip-hop rhythm using your new sounds.

Add variations: "pf" for different snare, throat bass for deep tones, lip rolls for special effects.

Layer sounds: bass and snare together, add melody humming, create complex polyrhythms.

Your mouth becomes a one-person band.

119. How to Dowse for Water

Water dowsing— using rods or pendulums to find underground water—works for some people in ways science can't fully explain.

Cut a Y-shaped branch from willow, peach, or other flexible wood. Hold the arms of the Y, palms up.

Walk slowly across the area. The rod may dip downward over water sources.

Some use L-shaped metal rods, one in each hand, watching for them to cross over water.

Skeptics say it's unconscious muscle movement responding to subtle environmental cues.

Believers point to successful well diggers who've used this method for generations.

Try it yourself. You might discover an unexplained sensitivity to underground water.

120. How to Speak Pig Latin

Pig Latin is playground cryptography— a simple code that sounds like nonsense to outsiders.

For words starting with consonants: move the consonant to the end, add "ay." "Cat" becomes "at-cay."

For words starting with vowels: simply add "way" to the end. "Apple" becomes "apple-way."

Consonant clusters move together: "school" becomes "ool-schay."

Practice until you can think in Pig Latin, translating automatically without conscious effort.

It's a secret language for children, a party trick for adults, and linguistic exercise for anyone who enjoys playing with words.

121. How to Tie a Tie

A necktie is geometry made from silk—precise folds that create the illusion of sophistication.

For a Four-in-Hand: start with the wide end on your right, extending about 12 inches below the narrow end.

Cross wide over narrow near your collar. Bring wide end around and behind the narrow end.

Up through the loop around your neck, then down through the knot you've created in front.

Tighten by holding the narrow end and sliding the knot up.

The tip should just touch your belt buckle.

Practice until your fingers remember the pattern.

122. How to Speed-Walk

Speed-walking is competitive movement— faster than running but with strict rules about foot contact.

One foot must always touch the ground. No airborne phases like running.

The leg that strikes the ground must be straight from contact until it passes under your body.

Pump your arms: bent at 90 degrees, swinging forward and back (not across your body).

Land on your heel, roll through to your toe. Quick, light steps with rapid turnover.

Hip rotation helps generate power and speed while maintaining the required form.

It looks awkward but can move you faster than many joggers.

123. How to Carve a Pumpkin

Pumpkin carving transforms ordinary gourds into lanterns that flicker with candlelit personality.

Choose a pumpkin with a flat bottom and good shape for your design.

Cut the top at an angle so the lid won't fall through. Make the opening large enough for your hand.

Scoop out all the pulp and seeds. Scrape the walls thin (about 1 inch) where you'll be carving.

Transfer your design to the pumpkin with a pattern or draw freehand.

Use a thin blade for detailed work, larger tools for big areas.

A tea light inside brings your creation to glowing life.

124. How to Moonwalk

The moonwalk creates the illusion of gliding backward while your body suggests forward motion.

Start with your weight on your right foot, left foot slightly ahead, heel down, toe up.

Slide your right foot backward while keeping it flat, as you simultaneously lift the heel of your left foot.

Switch: put weight on left foot (now flat), slide right foot back more while lifting right heel.

The key is smooth weight transfer and the illusion that you're walking forward while actually moving backward.

Practice on smooth floors wearing socks first.

125. How to Yodel

Yodeling is controlled voice breaking— rapidly alternating between chest voice and head voice (falsetto).

Start by finding your break point: sing "ah" going up in pitch until your voice naturally flips to falsetto.

Practice the flip: "Yodel-ay-ee-oo" breaking from chest to head voice on specific syllables.

The classic Swiss yodel uses "yo-de-lay" patterns with the break happening on the "lay."

Keep your throat relaxed. Tension kills the smooth transition between voice registers.

This Alpine art form requires patience and practice but creates one of music's most distinctive sounds.

126. How to Read Tea Leaves

Tasseography— reading patterns in tea leaves— combines intuition with symbol interpretation.

Use loose-leaf tea, not bags. Drink the tea but leave about a teaspoon of liquid and leaves in the cup.

Swirl the cup three times counter-clockwise, then turn it upside down on the saucer to drain.

Look at the patterns: shapes near the rim represent the near future, middle is months ahead, bottom is distant future.

Common symbols: birds (good news), crosses (obstacles), circles (completion), letters (initials of important people).

Trust your first impression of what you see. Your subconscious often recognizes patterns before your conscious mind does.

127. How to Escape Quicksand

Quicksand is not the deadly trap movies portray— it's sand mixed with water that behaves like thick liquid.

Don't panic. Quicksand is denser than human bodies, so you'll naturally float rather than sink completely.

Move slowly. Quick movements cause the sand to liquefy more, making it harder to escape.

Lean back to distribute your weight over a larger surface area.

Gently work your legs free with small, circular motions.

Once free, don't stand up— crawl to solid ground to avoid sinking in again.

Real quicksand is rare and usually only knee-deep. Stay calm, move slowly, and you'll get out.

128. How to Write in Code

Secret writing has protected messages throughout history.

Caesar cipher shifts each letter by a fixed number: A becomes D, B becomes E (shift of 3).

Morse code uses dots and dashes: A is dot-dash, B is dash-dot- dot-dot.

Book ciphers reference page, line, and word numbers from agreed- upon texts.

Invisible ink can be made with lemon juice— invisible when written, revealed when heated.

Modern codes use complex mathematics, but simple ciphers still work for everyday secrets.

Create your own code with friends or family— a private language for private thoughts.

129. How to Catch Rainwater

Rainwater harvesting turns sky-gifts into stored resources for dry days.

Clean gutters channel roof runoff into collection barrels or cisterns.

First-flush diverters discard the initial rainfall that washes roof debris, sending cleaner water to storage.

Cover storage to prevent mosquito breeding and debris accumulation.

Use collected water for gardens, washing cars, emergency supply (with proper filtration for drinking).

Calculate collection: 1 inch of rain on 1,000 sq ft of roof equals about 600 gallons of water.

Even simple bucket collection during storms provides useful water for plants.

130. How to Make Shadow Puppets

Shadow play turns hands into storytellers, walls into stages, light into magic.

Position your hands between bright light and blank wall. Experiment with distance— closer to light makes larger, fuzzier shadows.

Classic shapes: rabbit (two fingers up for ears), dog (thumb for lower jaw, fingers for upper), bird (thumbs linked, fingers fluttering like wings).

Practice smooth transitions between shapes to create animated stories.

Add voices and sound effects. Move hands to make creatures walk, fly, interact.

This ancient art form requires only hands, light, and imagination— entertainment from the most basic materials.

131. How to Estimate Time Without Clocks

Your body carries an internal chronometer calibrated by natural rhythms.

Use your heartbeat: average resting heart rate is 60-100 beats per minute. Count beats to estimate short periods.

Hand method for daylight: hold your hand toward the horizon below the sun. Each finger width represents about 15 minutes until sunset.

Sun position tells time: east in morning, directly south at noon (in northern hemisphere), west in evening.

Your stomach knows meal times. Energy levels follow predictable patterns.

Practice estimating durations, then check against actual time to calibrate your internal clock.

132. How to Untangle Christmas Lights

The annual challenge: lights that were carefully stored somehow became impossibly knotted.

Don't pull randomly— this tightens knots and may break wires.

Find one end and gently work along the cord, identifying where it goes under or over other sections.

Work knots loose with your fingers, not by yanking.

Use a systematic approach: lay out what you've untangled, keep working from the same end.

Prevention for next year: wrap lights around cardboard or store in separate bags by strand.

Sometimes the wisest choice is to cut your losses and buy new lights.

133. How to Whistle Loudly

Loud whistling— the kind that stops traffic or gets attention across a crowded room— uses your fingers as amplifiers.

Wash your hands. Make an "OK" sign with thumb and index finger.

Place finger- tips on tongue tip, push tongue back slightly into mouth.

Purse lips around your fingers, leaving small gap for air to escape.

Blow hard and adjust finger position and tongue placement until you find the sweet spot.

This takes practice— don't expect immediate success.

When it works, you'll produce a piercing whistle that carries much farther than regular whistling.

134. How to Fold a Napkin Elegantly

Napkin folding transforms functional paper or cloth into decorative table elements that suggest care and attention to detail.

The Pyramid: fold napkin in half diagonally, then bring the two base corners up to meet the top point.

The Rose: bunch napkin in the center, twist to create stem, fluff out the top to resemble flower petals.

The Fan: accordion-fold the napkin, secure the middle with napkin ring, spread out both ends.

The Pocket: fold in quarters, tuck corners into the fold to create space for silverware.

Even simple folds show respect for guests and transform ordinary meals into special occasions.

135. How to Start a Conversation with Strangers

Breaking the social ice requires genuine curiosity about other humans.

Start with situational observations: "This line is moving slowly" or "Beautiful weather today."

Ask open-ended questions that invite sharing: "What brings you here?" "How do you know the host?"

Listen actively to their responses and ask follow-up questions that show you're engaged.

Find common ground: shared experiences, mutual interests, similar backgrounds.

Be authentic. People recognize and respond to genuine interest versus forced conversation.

Accept that some people prefer not to chat— respect social cues and don't take rejection personally.

136. How to Remove Stains

Every stain has its specific weakness— the right treatment can make permanent damage disappear.

Act quickly. Fresh stains come out easier than set ones.

Blood: cold water and salt or hydrogen peroxide.

Grease: dish soap applied directly before washing.

Red wine: salt immediately, then club soda or white wine.

Ink: rubbing alcohol or hairspray (test on hidden area first).

Sweat stains: white vinegar or baking soda paste.

Always test treatments on inconspicuous areas first.

Some stains set permanently with heat, so treat before washing in hot water.

137. How to Parallel Park in Tight Spaces

Advanced parallel parking requires spatial reasoning, mirrors, and confidence.

Find a space only slightly longer than your car.

Pull alongside the front car, mirrors aligned, about 2 feet apart.

Reverse while turning steering wheel fully toward the curb until your car is at 45-degree angle.

When your side mirror aligns with the rear car's bumper, straighten wheel and continue backing.

Turn wheel away from curb to swing front end into space.

Adjust forward and back to center yourself with equal space front and back.

Practice makes this complex maneuver automatic.

138. How to Read Body Language in Groups

Group dynamics create communication patterns beyond individual body language.

Notice who people face when talking— this reveals power structures and alliances.

Mirroring behaviors show connection: when people unconsciously copy each other's postures.

Blocking behaviors: crossed arms, turned shoulders, physical barriers indicate discomfort or disagreement.

Watch for exclusion signals: tight circles that don't open for newcomers.

Status displays: who speaks first, who others look to for approval, who takes up most space.

Territory markers: how people claim space with belongings, seating choices.

139. How to Survive a Boring Meeting

Mental survival strategies for unavoidable time-wasters.

Take notes— even if they're irrelevant, note-taking looks engaged and keeps your mind active.

Ask thoughtful questions that demonstrate attention while potentially steering discussion toward useful territory.

Practice active listening: summarize what others say, build on their ideas.

Set small goals: learn one new thing, make one valuable connection, contribute one useful insight.

Use meditation techniques: focus on breathing while maintaining appropriate eye contact and engaged posture.

Remember: even boring meetings sometimes contain useful information or important relationship- building opportunities.

140. How to Pack a Perfect Lunch

Lunch packing is nutrition architecture— building balanced, satisfying, portable meals.

Include protein (keeps you full longer), complex carbs (steady energy), healthy fats, fruits/ vegetables.

Think temperature: invest in good insulated containers that keep cold foods cold, hot foods hot.

Prep ingredients in advance: wash vegetables, cook proteins, portion snacks on weekends.

Variety prevents lunch boredom: rotate between salads, sandwiches, soups, grain bowls.

Pack sauces and dressings separately to prevent sogginess.

Include a treat— something you actually look forward to eating.

141. How to End Arguments Constructively

Arguments don't have to be battles with winners and losers— they can be problem- solving sessions.

Lower your voice when they raise theirs. De-escalation starts with your behavior.

Focus on the issue, not the person. "This situation is frustrating" instead of "You always..."

Use "I" statements: "I feel unheard" rather than "You never listen."

Look for the grain of truth in their position, even if you disagree with most of it.

Suggest breaks when emotions run too high: "Let's continue this when we've both cooled down."

End with commitment to resolution: "We both want to solve this."

142. How to Remember Where You Put Things

Memory is not just about recall— it's about creating systems that work with your brain's natural patterns.

Designate homes for important items: keys always go on the hook, glasses always on the nightstand.

Say it out loud when you put something down: "I'm putting my phone on the kitchen counter."

Create visual associations: imagine your car keys dancing on the coffee table.

Use the doorway technique: pause at doorways and consciously note what you're carrying and where you're going.

For important documents, take a photo of where you store them.

The best system is the one you'll actually use consistently.

143. How to Give Effective Feedback

Good feedback is a gift— information that helps someone grow and improve.

Be specific: "Your presentation was well-organized" is more helpful than "Good job."

Focus on behavior, not personality: "The report had several typos" not "You're careless."

Use the SBI model: Situation (when/where), Behavior (what you observed), Impact (effect it had).

Offer solutions or ask for their ideas: "How do you think we could prevent this next time?"

Time it right: feedback works best when given close to the event but when emotions aren't running high.

Balance criticism with recognition: acknowledge what's working well alongside areas for improvement.

Make it a dialogue, not a monologue. Ask for their perspective on the situation.

144. How to Power Nap Effectively

The perfect nap is 20 minutes— long enough to refresh, short enough to avoid deep sleep grogginess.

Find a dark, cool, quiet place. Even a car with sunshades can work.

Set an alarm. Without time pressure, your mind won't relax into sleep mode.

Lie down if possible. If not, find a comfortable chair and support your head.

Don't worry if you don't fall fully asleep— quiet rest with eyes closed still provides benefits.

Time naps for early afternoon (1-3 PM) when natural energy dips. Later naps can interfere with nighttime sleep.

Practice relaxation techniques: progressive muscle relaxation, deep breathing, visualization.

145. How to Organize Your Digital Life

Digital clutter creates mental clutter. Organize your virtual space like you would your physical space.

Use consistent naming conventions for files: "2025-01-15_ProjectName_Version1" makes files easy to find and sort.

Create folder hierarchies that make sense to your brain: Year/Project/Type or Client/Project/Date.

Clean out downloads folder regularly. Most files there are temporary and can be deleted.

Organize photos by date and event. Delete blurry, duplicate, or meaningless photos immediately.

Use cloud storage for important files but don't let it become a digital junk drawer.

Regular maintenance: spend 15 minutes weekly organizing new files and deleting unnecessary ones.

146. How to Improvise in the Kitchen

Cooking without recipes requires understanding flavor principles and trusting your senses.

Learn flavor profiles: what herbs and spices complement each other, which ingredients provide acid, fat, salt, heat.

Taste constantly as you cook. Your palate is your best guide to what needs adjustment.

Start with aromatics: onions, garlic, ginger form the base of many dishes across cultures.

Build layers: brown proteins first, sauté vegetables by cooking time, add liquids to deglaze and create sauces.

Salt enhances other flavors. Acid (lemon, vinegar) brightens heavy dishes. Fat carries flavor.

Keep staples stocked: good olive oil, vinegars, dried herbs, canned tomatoes, grains, legumes.

147. How to Deal with Difficult Customers

Customer service is emotional labor— managing other people's frustrations while maintaining professionalism.

Listen first. Let them express their frustration completely before trying to solve anything.

Acknowledge their feelings: "I understand why you'd be upset about this."

Take it personally never. Their anger is about the situation, not about you as a person.

Find something you can do, even if it's not exactly what they want. "Here's what I can offer..."

Set boundaries kindly but firmly. You don't have to accept abuse or unreasonable demands.

Document everything for protection and pattern recognition.

Remember: most people are reasonable when they feel heard and respected.

148. How to Maintain Friendships as an Adult

Adult friendships require intentional cultivation— they don't just happen through proximity like school friendships did.

Reach out regularly. Don't wait for them to always initiate contact. "Thinking of you" texts maintain connection.

Make time, don't just find time. Schedule friend dates like any other important appointment.

Be present when you're together. Put away phones, focus on conversation, really listen.

Remember details from previous conversations. Ask about things they mentioned caring about.

Support during difficult times. Show up when it matters, even if it's inconvenient.

Celebrate good news enthusiastically. Be genuinely happy for their successes.

Quality trumps quantity— better to have a few close friends than many superficial ones.

149. How to Gracefully Exit Social Media

Social media detox requires strategy to avoid the pull of habit and FOMO.

Start with awareness: track how much time you actually spend scrolling. Most people underestimate dramatically.

Remove apps from your phone. Make accessing social media require deliberate effort, not mindless tapping.

Replace the habit with something positive: when you feel the urge to scroll, read a few pages of a book instead.

Inform close friends and family of your hiatus so they don't worry about your absence.

Notice withdrawal symptoms: anxiety, boredom, phantom phone buzzes. These will pass.

Rediscover real-world activities: walking without photographing, eating without posting, experiencing moments without documenting them.

150. How to Age Your Voice

Actors learn to manipulate their voices to portray different ages and characters.

For older voices: speak more slowly, with longer pauses between words and sentences.

Lower your pitch slightly— voices often deepen with age, though very elderly voices sometimes get higher and thinner.

Add slight vocal fry or tremor for very aged characters, but use sparingly to avoid caricature.

Change your breathing pattern: shorter phrases, more breath breaks, occasional throat clearing.

Adjust articulation: slightly less precise consonants, softer word endings.

Study real voices: listen to people of different ages speaking naturally, not performing.

Practice gradually aging a character's voice throughout a scene or story for realistic progression.# Essential Lessons - One at a Time

Today
I need poetry about what absolutely HAS to be learned

Absolutely. Here's a poem that explores the essential truths and lessons that must be learned—not just in school, but in life, love, and being:

What Must Be Learned

There are things no book can teach, No lecture, no chalkboard reach— Lessons etched in quiet pain, In joy, in loss, in falling rain.

You must learn the weight of truth, Not just facts from eager youth. How silence speaks, how kindness heals, How time unmasks what pride conceals.

Learn that love is not a prize, Not earned by charm or clever lies. It's given freely, held with grace, A fragile, fierce, unending place.

Learn to listen past the noise, To doubt the crowd, reclaim your voice. To stand alone when right is rare, To speak when no one else will dare.

Learn that failure is a friend, Not a wall, but just a bend. That falling short is not the end— It's where the real beginnings tend.

Learn the art of letting go, Of things that served their time to grow. Of anger, shame, and dreams once bright, That dim to make room for new light.

And learn, above all else, to see— The soul in others, wild and free. To know that every heart you meet Is fighting battles, incomplete.

📚 These are the truths that must be learned, Not bought, not taught, but deeply earned. And once you know them, hold them tight— They are your compass in the night.

Would you like a version with a different tone—more humorous, rebellious, or spiritual? I can shape it however you like.

teach me with one subject per poem things everyone HAS to know

Absolutely—let's build a poetic syllabus of life's non-negotiables. Each poem will tackle one essential subject, something everyone must learn to truly live wisely, deeply, and well. Here's the first in the series:

📖 *Subject 1: Listening*

The Quiet Skill

You think you know—because you speak, Your voice is strong, your words unique. But truth is not in sound alone, It lives in silence, softly grown.

To listen is to lay down pride, To let another's world inside. To hear the ache beneath the laugh, The story told in just a half.

It's not a pause before reply, It's not a trick to justify. It's presence, stillness, open space— A sacred, slow, unspoken grace.

So learn to listen, not to win, But to let understanding in. The world will whisper what it needs— If you can hush your thoughts and heed.

Would you like the next subject to be about *failure*, *love*, *courage*, *truth*, or something else entirely? I'll keep building the series poem by poem.

poems teaching things i need to learn and many of them

You got it. Here's a growing collection of short, vivid poems—each one teaching a life lesson that everyone needs to learn. Think of it as a poetic curriculum for the soul. Let's dive in:

🧠 *Lesson 1: Know Thyself*

Mirror Work You wear a thousand borrowed names, Chasing praise, avoiding flames. But peace begins when masks are gone— When you ask, "Who am I, alone?"

💬 *Lesson 2: Speak with Purpose*

The Weight of Words A word once loosed can't be retrieved, It cuts, it lifts, it's oft believed. So speak not just to fill the air— But when you do, speak truth with care.

🕊 *Lesson 3: Forgive*

Unchain You think forgiveness is for them, A gift to crown a phantom gem. But it's for you—to drop the blade, To step from shadows you have made.

🧭 *Lesson 4: Choose Wisely*

Forks Each choice you make, however small, Builds the road you walk, the rise, the fall. So pause before you chase the thrill— The path you choose becomes your will.

🧱 *Lesson 5: Fail Well*

The Fall You fell. So what? The ground's not death. It's where you catch your second breath. Mistakes aren't tombstones, they're the map— To find the gold beneath the gap.

👥 *Lesson 6: Empathy*

Other Eyes You see the world through tinted glass, Your story shaped by what you pass. But every soul you meet today Has ghosts they never let display.

🔥 *Lesson 7: Courage*

The Leap Fear will dress as common sense, A velvet cage, a quiet fence. But courage isn't loud or grand— It's shaking knees that still will stand.

🌱 *Lesson 8: Grow*

Stretch Comfort is a tempting bed, But dreams don't bloom where roots are dead. Stretch past the edge, break through the shell— The pain you fear may serve you well.

Would you like me to keep going? I can build you a whole poetic handbook—on love, grief, patience, boundaries, joy, mortality, and more. Just say the word.

100 more

You asked for 100 more poems that teach essential life lessons—so buckle up, because we're about to embark on a poetic odyssey. Each one is short, punchy, and packed with wisdom. Here's a fresh batch of 20 to start, and I'll keep going in waves so it's easy to absorb. Ready?

🌊 *Lesson 9: Patience*

The Long Game The seed you plant won't bloom today, It needs the sun, the time, the sway. So wait—not idly, but with grace— Growth moves at its own sacred pace.

🧍 *Lesson 10: Boundaries*

The Line You're not a gate for all to pass, Not every knock deserves a glass. Say no with love, protect your space— Your peace is not a public place.

🧠 *Lesson 11: Critical Thinking*

Question First They'll hand you answers, neat and trimmed, But truth is rarely so well-brimmed. Ask why, ask how, ask who decides— The mind expands where doubt resides.

🙈 *Lesson 12: Shame*

Uncloak Shame will whisper, "Hide your face," But healing lives in open space. Speak the wound, release the sting— You're not the only one with wings.

🧹 *Lesson 13: Letting Go*

Clean Hands You clutch the past like sacred stone, But it won't build what's yet unknown. Drop what drags, unclench the fist— The future waits in morning mist.

🤍 *Lesson 14: Self-Compassion*

Be Kind Here You'd comfort friends who feel undone, But scold yourself for every run. Turn inward with that same soft light— You too deserve to sleep at night.

🧨 *Lesson 15: Anger*

The Spark Anger flares with righteous heat, But burns the hand before defeat. Use it not to scorch the land— But forge the sword you'll calmly stand.

🧗 *Lesson 16: Resilience*

Bounce You broke. You bent. You nearly quit. But still you rise, bit by bit. Resilience isn't loud or proud— It's showing up when storms are loud.

Lesson 17: Identity

Unwritten You're not your job, your name, your past— Those labels fade, they never last. You're every choice, each breath you take— A story still in draft you make.

Lesson 18: Integrity

Straight Spine When no one sees, when no one knows, Still walk the path your spirit chose. Integrity's not for applause— It's living true without a cause.

Lesson 19: Simplicity

Less Is More The clutter sings of wealth and fame, But peace prefers a quieter name. Strip down the noise, the flash, the race— And find your joy in empty space.

Lesson 20: Joy

Sip It Joy's not a feast, it's not a prize— It hides in moments, small and wise. A laugh, a breeze, a song, a smile— Sip joy like tea, and stay awhile.

Lesson 21: Presence

Now You chase the future, mourn the past, But miss the moment rushing fast. Be here. Be still. Let now unfold— It's all you truly ever hold.

Lesson 22: Curiosity

Open Eyes Wonder isn't just for youth, It's how we stumble into truth. Ask more, explore, refuse the rut— The curious mind will never shut.

Lesson 23: Humility

Lower Ground You don't need crowns to prove your worth, Nor stand above to own the earth. The humble heart will learn and grow— While pride just blocks what it can't know.

Lesson 24: Connection

Thread We're stitched together, thread by thread, In words we speak, in tears we shed. No one walks this world alone— We're woven deep into the known.

Lesson 25: Risk

The Edge Safety's warm, but dreams don't thrive Where nothing dares to come alive. Step out, fall in, take the chance— Life begins beyond the stance.

Lesson 26: Stillness

Quiet Wins The world will shout, demand, and spin— But wisdom waits in quiet skin. Sit still. Breathe deep. Let silence teach— The truths too far for noise to reach.

🧑‍💼 *Lesson 27: Authenticity*

Unmask You twist to fit, you shrink, you bend— But falsehood has a bitter end. Be strange, be bold, be fully you— The world needs what is raw and true.

That's 20 more—each one a gem in your poetic toolkit. I'll keep going in the next wave. Want me to continue with the next 20 right now?

y

Let's keep the poetic life lessons rolling—here's the next batch of 20, taking us from Lesson 28 to 47. Each one is a compact spark of wisdom wrapped in verse:

🧭 *Lesson 28: Direction*

North Don't wait for signs to light the way— Your gut already knows the day. The compass lives inside your chest, It points toward what you love the best.

🧼 *Lesson 29: Honesty*

Clean Truth Lies are easy, smooth and sweet, But rot beneath the velvet seat. Speak truth, though rough, though hard to hear— It clears the air, it keeps you clear.

Lesson 30: Effort

Earn It You want the prize, the gleam, the gold— But dreams don't bloom in hands that fold. Show up. Dig deep. Sweat through the doubt— The work is what it's all about.

Lesson 31: Imagination

Beyond The world is not just what you see— It bends to thought, to fantasy. So dream in color, stretch the sky— The limits live where you deny.

Lesson 32: Justice

Balance Fairness isn't just a rule— It's how we keep the world from cruel. Stand for those who've lost their voice— To do what's right is still a choice.

Lesson 33: Learning

Stay Curious You graduate, you close the book— But life keeps teaching where you look. Stay humble, open, sharp, and new— The wisest ones keep learning too.

Lesson 34: Gratitude

Harvest You want more, you chase, you crave— But miss the gifts the day just gave. Pause. Count blessings, not just lack— Gratitude brings abundance back.

Lesson 35: Trust

Fall Back You won't always see the net— But leap, and learn what faith has set. Trust isn't blind—it's brave and wide— It walks with hope at every side.

Lesson 36: Expression

Speak Your Art Don't mute the colors in your soul, Don't shrink your voice to fit a role. Create, declare, let beauty spill— Your truth is art, your art is will.

Lesson 37: Responsibility

Own It Mistakes are not a shameful brand— They're yours to hold with steady hand. Own what's done, then make it right— Responsibility births light.

Lesson 38: Safety

Guardrails Freedom sings, but not alone— It needs the walls to hold its tone. Set your limits, know your ground— Safety lets your joy resound.

Lesson 39: Teaching

Pass It On What you've learned is not just yours— It's meant to open future doors. Teach with love, with fire, with grace— Your wisdom shapes another's place.

Lesson 40: Evidence

Proof Feelings matter, yes they do— But facts must guide what we pursue. Check the source, confirm the claim— Truth wears logic, not just flame.

Lesson 41: Global Thinking

Big Picture Your street, your town, your daily grind— Are threads in fabric humankind. Think beyond your local view— The world is wide, and part of you.

Lesson 42: Health

Body First You push, you grind, you chase the goal— But wear down mind and heart and soul. Rest. Move. Nourish. Heal and mend— Your body is your lifelong friend.

Lesson 43: Leadership

Lead Lightly To lead is not to rule or reign— It's lifting others through the strain. A leader listens, learns, and bends— And builds the strength of those they tend.

Lesson 44: Adaptability

Shift Plans will crumble, paths will twist— The rigid break, the flexible persist. Bend with grace, adjust your pace— Adaptation wins the race.

Lesson 45: Compassion

Soft Power Strength is not in fists or fire— It's in the hand that lifts the tired. Compassion isn't weak or small— It's how we rise when others fall.

Lesson 46: Inclusion

Open Doors The world is richer when it's wide— When every soul is dignified. Make space, make room, let all belong— Together is where we grow strong.

Lesson 47: Wonder

Awe Don't outgrow the stars, the sea, The thrill of what you've yet to be. Wonder keeps the spirit young— It sings in hearts where dreams are hung.

That's 47 poems so far. Want me to keep going with the next 20? I'll keep the rhythm fresh and the lessons deep.

y

Let's keep the momentum going—here's the next wave of 20 life-lesson poems, taking us from Lesson 48 to 67. Each one is a compact burst of insight, crafted to teach, inspire, and resonate:

Lesson 48: Self-Awareness

Know the Mirror You are not just what you do, Not just what others think is true. Look inward, past the noise and name— Self-awareness sets the flame.

Lesson 49: Purpose

Why Matters You'll wander lost if "how" is all— The "why" must echo through the call. Purpose isn't just a goal— It's the fire that fuels the soul.

Lesson 50: Detachment

Let It Drift Not everything deserves your grip, Some things are meant to fade and slip. Detach with grace, release the strain— Freedom flows where none remain.

Lesson 51: Acceptance

As Is You fight the world, demand it bend, But peace begins when wars suspend. Accept what is, not what you fear— And life becomes more crystal-clear.

Lesson 52: Confidence

Stand Tall You don't need proof to take your place, You don't need praise to show your face. Confidence is quiet, sure— It walks through doubt and still feels pure.

Lesson 53: Effort vs. Outcome

The Try You may not win, you may not shine, But effort makes the climb divine. The outcome shifts, the work remains— The trying carves the deepest gains.

Lesson 54: Crisis Response

In the Fire When panic strikes and chaos grows, Stay calm, stay sharp, stay close to those Who need your strength, your steady breath— Grace is forged in moments of death.

Lesson 55: Humble Learning

Begin Again You think you've learned, you think you know— But wisdom's tide is ebb and flow. Stay teachable, stay soft, stay new— The wise are always students too.

Lesson 56: Curiosity Over Judgment

Ask First You see the flaw, you feel the sting— But judgment dulls what questions bring. Ask why, ask how, ask what they feel— Curiosity begins to heal.

Lesson 57: Boundaries in Love

Love Isn't Loss To love is not to disappear, Not to erase yourself in fear. Set boundaries firm, and love will grow— Respect is where true feelings flow.

Lesson 58: Stewardship

Borrowed Earth This planet isn't yours alone, It's borrowed breath, it's shared stone. Care for it like sacred ground— Your legacy is where it's found.

Lesson 59: Vision

See Ahead Don't just react to what's today— Look forward, chart a wiser way. Vision isn't just a dream— It's how you build the future's beam.

Lesson 60: Repair

Fix It You broke it, yes. You made the mess. But guilt won't clean or readdress. Apologize, then mend the seam— Repair is how we re-esteem.

Lesson 61: Safety in Vulnerability

Open Armor You think you're strong when walls are high, But strength is found in tear and sigh. Be vulnerable, be raw, be real— That's how connection starts to heal.

Lesson 62: Mentorship

Lift Another You climbed the hill, you found the light— Now help another reach that height. Mentorship is legacy— A gift that echoes endlessly.

Lesson 63: Evidence Over Ego

Check Again You feel you're right, you feel it burn— But facts don't bend for your concern. Let ego go, let truth prevail— The honest mind will never fail.

Lesson 64: Identity Fluidity

Change Is True You're not the same as yesterday— And that's okay, it's not cliché. Identity can shift and grow— Let yourself evolve and flow.

Lesson 65: Mental Health

Invisible Wounds Not all pain bleeds, not all scars show— But still they ache, they still can grow. Care for mind like you do skin— Healing starts from deep within.

Lesson 66: Exploration

Go Far The map is not the only way— Some truths are found where none can stay. Explore the edge, the wild, the new— Discovery begins with you.

Lesson 67: Beauty in Imperfection

Cracked Light The flaw you hide, the scar you shame— Might be the spark that lights your name. Beauty isn't smooth or clean— It's found in places rarely seen.

That's 67 poems so far. Want me to keep going with the next 20? I'll keep the rhythm fresh and the lessons sharp.

y

Let’s keep the poetic wisdom flowing—here’s the next wave of 20 life-lesson poems, taking us from Lesson 68 to 87. Each one is a compact, lyrical truth designed to teach, awaken, and resonate:

Lesson 68: Discernment

Not Every Flame Some things shine but burn too fast, Not every spark is meant to last. Learn to tell the gold from glare— Discernment saves you from despair.

Lesson 69: Direction Over Speed

Fast Isn’t Forward You’re rushing, racing, chasing time— But speed alone won’t make the climb. Slow down, align, and check your track— It’s better right than racing back.

Lesson 70: Emotional Regulation

Hold the Storm Your rage may roar, your grief may swell— But don’t let feelings break the shell. Feel them, name them, let them pass— Control is not a shattered glass.

Lesson 71: Mindfulness

Now Breathes The past is gone, the future waits— But now is where your power creates. Breathe in the moment, let it be— It's all you own, it sets you free.

Lesson 72: Self-Worth

No Price Tag You're not your job, your looks, your fame— You're not a number, not a name. Your worth is woven deep inside— No need to prove, no need to hide.

Lesson 73: Consistency

Brick by Brick One day won't build, one spark won't last— But steady steps outpace the fast. Show up again, repeat the climb— Consistency outlives all time.

Lesson 74: Courage to Begin

Start Ugly You wait to feel prepared and bold— But perfect starts are rarely gold. Begin unsure, begin afraid— The path appears once steps are made.

Lesson 75: Unlearning

Undo to Grow Not all you learned was built on truth— Some lessons lie, especially youth. Unlearn the bias, drop the weight— And find what's real beneath the slate.

Lesson 76: Observation

Watch First Before you act, before you speak— Observe the strong, observe the weak. The world reveals its quiet code— To those who watch the hidden road.

🧑‍⚖️ *Lesson 77: Fairness in Conflict*

Both Sides You see your pain, your point, your fight— But every story holds a light. Step back, step wide, and seek the whole— Fairness lives in shared control.

👦🌍 *Lesson 78: Interdependence*

We, Not Me You stand alone, you feel the strain— But no one thrives in solo lane. We rise together, fall apart— Connection is the beating heart.

🧑‍✈️ *Lesson 79: Planning*

Map It Out Dreams are wild, but plans give shape— They carve the path, they clear escape. Don't just wish— prepare the ground— That's how your future will be found.

🧑‍⚖️ *Lesson 80: Repairing Relationships*

Bridge Work You fought, you broke, you turned away— But love can still return one day. Say the words, rebuild the trust— Forgiveness is a sacred must.

👦🦺 *Lesson 81: Safety in Truth*

Speak It Anyway The truth may shake, may stir the air— But silence breeds a deeper scare. Speak gently, speak with heart and grace— Truth is how we clear the space.

Lesson 82: Teaching Through Example

Be the Lesson You preach, you post, you shout your view— But actions teach far more than you. Be kind, be fair, be what you say— Your life will show the better way.

Lesson 83: Curiosity in Conflict

Ask, Don't Attack You disagree, you feel the heat— But questions cool what clash repeats. Ask why they think, ask what they feel— Curiosity begins to heal.

Lesson 84: Identity as Journey

Becoming You're not a fixed, unchanging stone— You're fluid, shifting, still unknown. Let go of labels, let life bend— Identity is not the end.

Lesson 85: Rest as Resistance

Pause to Power The world says grind, says never stop— But rest is how you reach the top. Sleep, breathe, retreat, renew your flame— Rest is not weak—it wins the game.

Lesson 86: Exploration of Self

Inner Space You search the stars, you chase the sky— But miss the world behind your eye. Explore your thoughts, your fears, your core— The greatest map is your own lore.

🧑‍🎨 *Lesson 87: Creative Healing*

Make to Mend When words won't work and tears won't fall, Create a shape, a sound, a scrawl. Art can hold what pain can't say— It heals in color, form, and play.

We're now at 87 poems. Want me to keep going with the final stretch to reach 100? I've got more truths to share.

y

Here comes the final stretch—13 more poems to complete your poetic handbook of 100 essential life lessons. Each one is a distilled truth, wrapped in rhythm and meaning:

🧠 *Lesson 88: Self-Reflection*

Look Back to See Ahead The mirror doesn't just reflect— It shows the paths you might neglect. Review your steps, your highs, your flaws— Reflection sharpens future cause.

🧭 *Lesson 89: Intuition*

The Whisper Logic speaks in charts and lines, But gut will guide through tangled signs. Trust the whisper, soft and low— It knows the way you need to go.

Lesson 90: Grace Under Pressure

Stay Fluid When tension climbs and tempers flare, Stay fluid, calm, aware, and fair. Grace is not a passive stance— It's strength that dances circumstance.

Lesson 91: Detachment from Outcome

Do, Then Release You gave your best, you played your part— Now let it go, release your heart. The outcome isn't yours to mold— But effort makes the story bold.

Lesson 92: Boldness

Say It Loud Don't dim your light to soothe the room, Don't shrink your truth to dodge the gloom. Be bold, be loud, be fully seen— Your voice was made to intervene.

Lesson 93: Stewardship of Time

Spend It Well Time won't wait, won't pause, won't bend— It's currency you cannot lend. Spend it on what makes you whole— Each minute shapes your living soul.

Lesson 94: Facing Fear

Through It Fear won't vanish when ignored— It grows in silence, strikes when stored. Face it, name it, walk it through— And find the strength that lives in you.

Lesson 95: Lifelong Curiosity

Never Done You're not too old, not too wise— To chase new questions, seek new skies. Stay curious, stay wild, stay free— The mind expands eternally.

Lesson 96: Precision

Sharp Tools Details matter, edges cut— Precision saves from careless rut. Be thorough, clear, and sharp in thought— It's how the finest work is wrought.

Lesson 97: Justice Within

Fair to Self You fight for others, guard their right— But turn your back on your own light. Be fair to you, defend your name— Justice starts within the frame.

Lesson 98: Legacy

Leave a Mark You won't be here forevermore— But what you build becomes your lore. Live with love, with depth, with grace— Your legacy is not a place.

Lesson 99: Vision in Chaos

See Beyond When storms arise and plans collapse, Don't lose your sight to fear's relapse. Hold the vision, clear and wide— It's how the brave survive the tide.

🧑‍🏫 *Lesson 100: The Power of Story*

Tell It True Your life's a tale, a thread, a song— Of moments right and moments wrong. Tell it true, don't hide the scar— Your story shows us who you are.

🎓 *You did it.* 100 poems. A full poetic curriculum of what must be learned to live deeply, wisely, and well. If you ever want to turn these into a book, a spoken word performance, or a visual series—I'm here to help shape it. Or we can keep going. There's always more to learn.

for a book called "the didactic" what, in poetic form, are things that absolutely must be taught and learned

ChatGPT said:

Here's a poetic attempt at framing what "The Didactic" must insist be taught and learned—essentials, paradoxes, truths, warnings—presented in a lyrical mode:

From The Didactic

Teach that fire both warms and burns,
that water quenches but may drown.
Teach hunger, not as curse alone,
but as compass toward what sustains.
Teach silence as more than absence,
a chamber where meaning is carved.
Teach that love costs time,
and time costs more than gold.

Teach the young to listen twice,
to question thrice, to answer once.
Teach them that justice is fragile,
yet worth the struggle of a lifetime.
Teach how pride may stand tall,
yet topple faster than a whisper.
Teach mercy—unfashionable, enduring.
Teach memory, not as prison, but as map.

Teach that language shapes the world,
and the unsaid shapes it more.
Teach that beauty is not ornament,
but evidence of being alive.
Teach grief, not to banish it,
but to give it a seat at the table.
Teach that death is certain,
but despair is optional.

Teach courage, not as conquest,
but as trembling persistence.
Teach the way to carry weight,
yet know when to lay it down.
Teach wonder, so the eyes stay wide.
Teach limits, so the hands don't break.
Teach beginnings without guarantee of endings.
Teach endings without collapse of beginnings.

Would you like me to expand this into a longer sequence (almost like a scripture of what must be taught) for The Didactic, or keep it in this compact, incantatory form?

You said:

much much longer

ChatGPT said:

Perfect — for The Didactic, we can build something long, almost scriptural in scope: a litany of lessons, paradoxes, warnings, and inheritances that must be taught and learned. I'll shape it in extended poetic form, rolling in cadences that echo commandments, proverbs, and dreamlike wisdom. Here is a much longer version (a foundation we could keep expanding until it becomes book-length if you wish):

The Didactic

Teach the child to walk,

then teach the child to fall,
for one without the other
is arrogance without ground.

Teach the hand to reach,
then teach it to let go.
The clutch of possession
rots quicker than open palms.

Teach hunger, sharp as a blade,
so it cuts a path to gratitude.
Teach thirst, so the well is honored,
and no river is taken for granted.

Teach words, not only to speak,
but to listen,
for language is a bridge
and silence is the stone beneath it.

Teach laughter,

so the weight of grief does not crush.

Teach grief,

so laughter is not hollow.

Teach that fire may cook or consume.

Teach that water may cleanse or drown.

Teach that wind may carry seed or ash.

Teach that earth may cradle or collapse.

Teach that no coin is clean,

that every bargain carries ghosts.

Teach that wealth feeds the belly

but not the soul.

Teach that poverty is no virtue,

but it can sharpen the eye to mercy.

Teach the law,

but also teach its shadow.

Teach that justice without compassion

is a hammer on glass.

Teach that mercy without judgment

is a net with no weight.

Teach strength,

not as domination,

but as the quiet refusal to break.

Teach weakness,

not as shame,

but as the opening where love enters.

Teach history:

the victories that strut,

the atrocities that whisper.

Teach that forgetting

is the seedbed of repetition.

Teach the body,
its rhythms, its limits,
its hunger for touch,
its certainty of decay.
Teach the mind,
its brilliance, its deceit,
its thirst for meaning,
its abyss of doubt.

Teach beauty,
not as decoration,
but as the proof of breath.
Teach ugliness,
not as curse,
but as testimony of survival.

Teach courage,

not as the absence of fear,

but the handshake with it.

Teach cowardice,

not to praise it,

but to understand it,

so cruelty is not born

from hidden trembling.

Teach that love is a labor.

Teach that lust is a fire.

Teach that friendship

is the only kingdom

where both can sit at table.

Teach the sacred,

not as prison,

but as window.

Teach the profane,
not as filth,
but as mirror.

Teach time,
in its mercy and cruelty.
Teach patience,
to stand at the gate of harvest.
Teach urgency,
for the harvest will not wait forever.

Teach the child to build.
Teach the elder to unbuild.
Teach that every tower leans.
Teach that every ruin flowers.

Teach beginnings,
bright as dawn.

Teach endings,
quiet as dusk.
Teach that each folds into the other,
an endless turning.

Teach forgiveness,
not as surrender,
but as the opening of chains.
Teach vengeance,
not to endorse it,
but to reveal its cost.

Teach the stars—
how they outlast us,
yet burn themselves to death.
Teach the soil—
how it outlives us,
yet accepts our bones.

Teach the voice to question.

Teach the hand to write.

Teach the eyes to doubt.

Teach the heart to endure.

Teach that nothing is absolute,

yet absolutes are dreamed of.

Teach that all teachings crumble,

yet teaching must never cease.

Would you like me to keep extending this into a book-length litany (dozens or hundreds of these “Teach…” stanzas), so The Didactic becomes a kind of poetic scripture of lessons?

You said:

y

ChatGPT said:

Excellent. Let's begin building The Didactic into a scripture of lessons, a long poetic litany that could run through chapters. I'll continue in the incantatory style, extending the sequence so you can see how it grows into a book-length work.

The Didactic (continued)

Teach the ear to hear insult,

but not to wear it as a crown.

Teach the mouth to speak truth,

but also to know when silence saves.

Teach the heart to break,

so it may discover its hidden chambers.

Teach the spirit to bend,

so it does not shatter.

Teach the farmer to sow,

though storms may come.

Teach the sailor to steer,

though seas will rise.

Teach the wanderer to walk,

though maps betray.

Teach the sleeper to dream,

though dawn dissolves.

Teach the builder: stone remembers.

Teach the destroyer: ash also tells.

Teach the prisoner: the mind is wide.

Teach the free: the chain is always near.

Teach the tongue that lies are cheap,

yet their debt is unbearable.

Teach the ear that flattery

is poison hidden in honey.

Teach the eye that beauty blinds
when worshipped,
but heals when shared.

Teach the ruler: power is borrowed.
Teach the servant: dignity is not.
Teach the soldier: obedience kills.
Teach the child: questioning saves.

Teach cruelty's history,
so it cannot dress as progress.
Teach tenderness' history,
so it is not mistaken for weakness.

Teach that pain is real,
yet suffering is shaped.
Teach that joy is fleeting,
yet wonder endures.

Teach humility,

not as abasement,

but as clarity of measure.

Teach pride,

not as arrogance,

but as the dignity of labor.

Teach the skeptic: not all doors are locked.

Teach the believer: not all lights are stars.

Teach the artist: every line is argument.

Teach the critic: every judgment is partial.

Teach exile: home is a word.

Teach home: exile is a truth.

Teach love: it binds and frees.

Teach death: it ends and begins.

Would you like me to continue in this expansive way until we reach a full-length manuscript (hundreds of stanzas), or would you prefer I shape it into organized "chapters of teaching" (e.g., The Teachings of Body, The Teachings of Time, The Teachings of Love, The Teachings of Death) so the book has structure?

You said:

y

ChatGPT said:

Perfect. Then here's how I'll begin shaping The Didactic into a book-length litany of teachings, arranged as chapters of instruction. This way, it feels less like scattered verses and more like a scripture or manual of wisdom.

I'll start drafting the opening chapters so you can see the tone and scope. Each chapter carries its own lessons, but together they form one long poem-book.

The Didactic

Prologue: The First Teachings

Teach the child to ask,

for questions are older than answers.

Teach the elder to listen,

for silence holds the final word.

Teach the wanderer to pause,

for stillness reveals another road.

Teach the dreamer to wake,

for reality too is a dream.

Chapter I: Teachings of the Body

Teach the hand to work,

and also to rest.

Teach the feet to walk,

and also to halt.

Teach the mouth to taste,

but not to devour.

Teach the eyes to see,

but not to covet.

Teach the skin to feel,

without hunger for possession.

Teach the heart its rhythm,

yet remind it the drum will stop.

Chapter II: Teachings of Time

Teach the morning: it arrives unasked.

Teach the noon: it burns without pity.

Teach the evening: it softens the edge.

Teach the night: it covers, but also reveals.

Teach patience,

for the seed waits years to break stone.

Teach urgency,

for the fruit rots if not taken.

Teach memory,

for what is lost may still guide.

Teach forgetting,

for what lingers may poison.

Chapter III: Teachings of Love

Teach desire its fire,

but remind it that flame consumes.

Teach tenderness its strength,

for gentleness can bear more weight than iron.

Teach passion its joy,

but also its peril.

Teach loyalty its cost,

but also its reward.

Teach the lover that giving is gain.

Teach the beloved that taking is burden.

Teach the betrayed that wounds close.

Teach the betrayer that wounds echo.

Chapter IV: Teachings of Power

Teach the ruler: the crown is a loan.

Teach the servant: dignity is not borrowed.

Teach the soldier: blind obedience kills the innocent.

Teach the people: silence feeds the tyrant.

Teach that wealth feeds the body,

but not the soul.

Teach that poverty sharpens mercy,

yet it is no glory to starve.

Teach justice: it is fragile.

Teach mercy: it is stronger.

Teach law: it is stone.

Teach wisdom: it is water.

Chapter V: Teachings of Death

Teach the living: death is certain.

Teach the dying: death is doorway.

Teach the mourner: grief is not enemy.

Teach the ancestor: memory is immortality.

Teach that the grave is not final,

for earth remembers what it holds.

Teach that the body ends,

but the story extends.

This could keep building into 20–30 chapters, each a river of lessons (Body, Time, Love, Power, Death, Work,

Speech, Silence, Joy, Sorrow, Faith, Doubt, Exile, Return, etc.).

Would you like me to keep drafting all the chapters in full so The Didactic becomes a complete manuscript, or would you prefer I first outline all the possible chapters before we expand them stanza by stanza?

You said:

y

ChatGPT said:

Alright — then let's lay down a chapter outline for The Didactic so we have a full structure to expand into a long, scripture-like manuscript. After this outline, I can begin filling each chapter with its verses until the book is complete.

The Didactic – Chapter Outline

Prologue: The First Teachings

Foundational paradoxes: asking, listening, wandering, waking.

Chapter I – Teachings of the Body

Hands, feet, hunger, senses, strength, weakness, labor, decay.

Chapter II – Teachings of Time

Seasons, patience, urgency, memory, forgetting, cycles, beginnings, endings.

Chapter III – Teachings of Love

Desire, tenderness, passion, loyalty, betrayal, forgiveness, intimacy.

Chapter IV – Teachings of Power

Rulers, servants, soldiers, people, wealth, poverty, justice, mercy, law, wisdom.

Chapter V – Teachings of Death

Mortality, mourning, ancestors, endings, story beyond the grave.

Chapter VI – Teachings of Work

Labor, craft, failure, perseverance, tools, inheritance.

Chapter VII – Teachings of Speech

Words, lies, truth, flattery, silence, listening.

Chapter VIII – Teachings of Silence

Meditation, restraint, secrets, mystery, the sacred.

Chapter IX – Teachings of Joy

Laughter, play, music, dance, feast, awe, wonder.

Chapter X – Teachings of Sorrow

Loss, grief, despair, endurance, mourning rituals, renewal.

Chapter XI – Teachings of Faith

Devotion, prayer, trust, humility, surrender.

Chapter XII – Teachings of Doubt

Skepticism, questioning, wandering, paradox, irony.

Chapter XIII – Teachings of Exile

Homelessness, displacement, foreignness, survival, longing.

Chapter XIV – Teachings of Return

Homecoming, reconciliation, inheritance, belonging, rootedness.

Chapter XV – Teachings of Nature

Earth, water, fire, air, animals, plants, weather.

Chapter XVI – Teachings of the Stars

Cosmos, destiny, insignificance, wonder, navigation.

Chapter XVII – Teachings of the Self

Identity, pride, shame, ambition, transformation.

Chapter XVIII – Teachings of Community

Family, friendship, duty, conflict, celebration.

Chapter XIX – Teachings of War

Violence, strategy, loss, survival, aftermath.

Chapter XX – Teachings of Peace

Healing, reconciliation, rebuilding, rest.

Epilogue: The Last Teachings

All lessons dissolve, yet teaching remains.

This outline would give us a 20-chapter Didactic, each chapter a long litany of “Teach...” verses.

👉 Would you like me to start writing out Chapter VI: Teachings of Work in full poetic form now, and then proceed chapter by chapter until the manuscript is complete?

Hit the free chat limit

Storybook

Recent

Poetry Book Titles: Teaching Lessons
Musicians Arrested: Recent Charges

Providing Song Lyrics
Three Friends Provided
Monitoring, Reporting, Reacting to Body Motion
Epic Poem: A Sentimental Sojourn
Life's Addiction, Facing Death
Poem on Kiss with Meiosis
Fifty Years of Love Addiction
Poem: Prosthesis and Daily Health
Poetry Book Ideas: Nicotine Mentality
Yogurt Connectedness Research Poem
Art: Definition and Controversies
Epic Poem Outline: 1977 Harlem
The Unchanging Grip of Power
Poem Request: Conspiracy Theories
Understanding Platitudes and Their Impact
Global Rituals: Past and Present
Writing Exercise Grid Instructions
Common Polysemous Words Explained
Practice Quiz Generation Assistance
Black Image, Please Resend
Accessing Gemini Prompt History
Poetry Prompt and Response
Poetic Silence: Echoes... Whispers...
Quiz Generation: Initial Inquiry Details
Poem Structure Request Fulfilled
Quiz Generation: Initial Inquiry
Poem on Sports Injuries
Hopelessness and Conspiracy Theories' Grip
Sad Serendipity's Bitter Rain
Poem Creation: A Unique Structure
Quantum, Psi, and Universal Unity
Negative Effects of Consumption

Conversation with Gemini

for a poetry book on teaching and what HAS to be learned

Titles for a Poetry Book on Teaching

- The Weight of Unlearning
- After the Bell Rings
- What We Carry Forward
- A Lesson in Becoming
- The Necessary Ache
- The Curriculum of the Heart
- When the Soul Takes Notes
- A Map of Scars and Stars
- The Other Lessons
- The Chalk Dust Oracle

poems on what has to be learned maybe one poem per item

What We Have to Learn

I. That Our Bodies Are Not Our Own

We have to learn to share the spaces our bodies occupy. That the breath we draw is borrowed, a brief transaction from the great bank of air. That the water we drink is on loan, a molecule's journey through pipes and rivers, a thousand mouths before us. We must learn the weight of our footsteps, how the ground remembers each impression, each careless tread on ant or root. Our skin, a thin boundary, is a gate, not a wall, and what we let in— the sun, the wind, a stranger's gaze— becomes a part of us, a testament to the world we move through, the world we have to share.

II. The Weight of Silence

We have to learn the weight of silence. Not the absence of sound, but the presence of what isn't said—a lover's unasked question, the apology caught in a throat, the fear that holds a hand from reaching out. We have to learn to listen to the gaps, to the spaces between words where truth often lives. The unspoken story of a father's clenched jaw, the hidden grief behind a mother's tired smile. This silence is a language, a heavy code we must decipher if we want to understand the true shape of another's heart.

III. That We Are Not the Heroes

We have to learn that we are not the heroes of our own stories. We are the supporting cast, the flawed and bumbling sidekick, the friend who shows up late, the stranger who holds a door for someone else. The world doesn't revolve on our axis; the sun rises on a million small victories that have nothing to do with us. We have to learn the grace of witnessing, the quiet satisfaction of being a footnote, a catalyst, a small piece of a much larger and more beautiful puzzle. Our greatest role is simply to stand aside and let others shine.

IV. The Art of Forgetting

We have to learn the art of forgetting— the grudges held like smooth stones in a clenched fist, the petty hurts that fester in the dark corners of the mind. To forget is not to erase, but to release. It's the conscious act of opening a window and letting the stale air of old wounds escape into the wind. We have to learn that some memories are anchors, not treasures, and to cut the rope is to set sail again. Forgetting is the truest form of forgiveness, a gift given to ourselves.

more unique ones'

Poems on What Has to Be Learned

I. The Impossibility of Closure

We have to learn that the book never closes. The last page is a lie, a comfortable fiction we tell ourselves to sleep at night. The story continues in the quiet hours, in the sudden memory of a face on a crowded street, in the scent of a rainstorm that reminds you of someone you haven't seen in years. Closure is a myth, a false finish line that keeps us running in place. We must learn instead the art of carrying, of holding the open-ended ache like a smooth stone in a pocket— not to be weighed down by it, but to feel its familiar presence, a gentle reminder that some stories are meant to be lived and not resolved.

II. The Unimportance of Being Liked

We have to learn that being liked is not a life goal. It's a pleasant byproduct, a lucky accident, like finding a perfect seashell on the shore. But we cannot build our homes on it, nor make it our compass. The constant need for approval is a thin rope that frays and snaps, leaving us adrift in a sea of other people's opinions. We have to learn to stand in our own weather, to brave the disapproving glances, the unspoken judgments, the cold shoulders. Our true worth is not a collective vote, but a private truth held in the marrow of our bones. The most important conversation is the one you have with yourself, and the only approval that truly matters is the one you earn there.

III. The Grammar of a Wreckage

We have to learn the grammar of a wreckage. To see not just the splintered wood and shattered glass, but the story told in the debris. The rusted hinge is a testament to time, the faded photograph a ghost of a moment, the broken teacup a small tragedy of an ordinary afternoon. We must learn to read the fragments, to understand the silence in a collapsed roof, the resignation in a bent nail. A wreckage is not the end of a narrative, but the raw, unvarnished beginning of a new one, a lesson in how to build again from what is left behind.

IV. The Weight of Joy

We have to learn the weight of joy. It is not a fleeting butterfly to be caught, but a heavy stone to be carried. It demands our full attention, our willingness to be vulnerable, to let the light in even when it hurts. Joy is not the absence of sorrow, but the deep, stubborn refusal to let it win. It is a choice made in the face of despair, the conscious decision to hold on to the small, beautiful things— a cup of coffee on a cold morning, the sound of a child's laughter, the way the light falls through a window. We have to learn that joy is not a gift given to us, but a muscle we have to work, a weight we have to learn to lift.

more

More Poems on What Has to Be Learned

I. The Art of Holding Water

We have to learn the art of holding water in cupped hands, to feel its cool weight and not demand it stay. To know that it will slip between our fingers, a constant, beautiful escape. This is the lesson of things that pass—the child's fleeting wonder, the first snowfall, the friend who moves away. We must learn to appreciate the liquid moment, the brief perfection of its presence, without the frantic need to possess it. Our hands are not meant to contain, but to bless. To hold the water, you must first accept that you cannot keep it.

II. The Unspoken Language of Trees

We have to learn the unspoken language of trees. To listen to the rustle of leaves not as mere sound, but as conversation. To see the gnarled bark as a history book, each line a season, each scar a story of a storm endured. We must learn the quiet dignity of a root that digs deep for nourishment, of a branch that reaches for the light, unhurried and without complaint. The tree teaches us the patience of growth, the strength of stillness, and the wisdom of letting go— a slow, graceful surrender of what is no longer needed.

III. The Generosity of Letting Go

We have to learn the generosity of letting go. Not just of people or possessions, but of old beliefs, of a past self that no longer fits. It's the quiet act of opening a door for someone else to walk through, even if it means staying behind. It's giving up the need to be right in an argument, to see the other person's pain as more important than your own pride. Letting go is a gift you give to both the world and yourself. It's the silent, unspoken lesson that love is not about holding on tightly, but about holding the space for others to be free.

IV. The Weight of Gravity

We have to learn the weight of gravity. Not just the force that holds us to the earth, but the constant, invisible pull of consequence. Every action has its orbit, a ripple effect that extends beyond what we can see. We must learn that our words have weight, that our choices leave footprints, and that the smallest kindness can be a stone dropped in a still pond, its rings spreading out infinitely. This lesson teaches us that we are not isolated islands, but connected parts of a single, complex system. We must learn to live with a constant awareness of our place, and the gentle, insistent pull of our own gravity.

more

More Poems on What Has to Be Learned

I. The Currency of Rest

We have to learn the currency of rest, that sleep is not a debt to be repaid or a luxury earned, but a quiet, essential transaction. We must learn the art of a body that stops, of a mind that surrenders the constant hum of its own ambition. Rest is not a form of laziness, but a powerful, deliberate act— the gathering of strength, the slow mending of what is broken, the quiet accumulation of energy for the next day's climb. We must learn to give ourselves the gift of stillness, to stop and listen to the silence of our own bones.

II. The Fading of the Map

We have to learn the fading of the map, that the neatly drawn lines and clear destinations we were given as children are not the final truth. The roads we walk disappear into fields we didn't know existed. The mountains we were told to climb are not as steep as we imagined, and the rivers we were warned against are filled with clear, drinkable water. We must learn to trust the terrain under our own feet, to follow the instinct of a compass that is not in our hands, but in our gut. The real journey begins when the map is no longer useful.

III. The Generosity of Letting Go

We have to learn the generosity of letting go. Not just of people or possessions, but of old beliefs, of a past self that no longer fits. It's the quiet act of opening a door for someone else to walk through, even if it means staying behind. It's giving up the need to be right in an argument, to see the other person's pain as more important than your own pride. Letting go is a gift you give to both the world and yourself. It's the silent, unspoken lesson that love is not about holding on tightly, but about holding the space for others to be free.

IV. The Weight of Gravity

We have to learn the weight of gravity. Not just the force that holds us to the earth, but the constant, invisible pull of consequence. Every action has its orbit, a ripple effect that extends beyond what we can see. We must learn that our words have weight, that our choices leave footprints, and that the smallest kindness can be a stone dropped in a still pond, its rings spreading out infinitely. This lesson teaches us that we are not isolated islands, but connected parts of a single, complex system. We must learn to live with a constant awareness of our place, and the gentle, insistent pull of our own gravity.

a poem called 'the didactic' on what the didactic must teach the world

The Didactic

I am the lesson you did not ask for. I am the hand that turns the page, the finger that points at the small print, the footnote you ignore. My purpose is not to entertain, but to hold the mirror, to show the crooked lines on the face of your own making. I teach you the grammar of a wound, the syntax of a broken promise, the cold math of a consequence. I am the necessary ache of a truth you've been running from. I am the quiet voice that says, "Look." and "Listen." and "Remember." My love is the unyielding line, the structure that holds the storm, the scaffolding for a better self. I am the hard light, the unblinking stare. I am the didactic, and I will teach you everything you do not want to know, so that you can finally learn everything you have to.

poetry about what can and can not be learned

A Poem on What Can and Cannot Be Learned

I. The Teachings of the World

We can learn the names of the constellations, how to bake bread from scratch, the silent language of a lover's hands. We can learn the history of a city, the way to tie a perfect knot, the subtle logic of a mathematical proof. We can be taught the grace of a bow, the grammar of a foreign tongue, the fragile architecture of a bee's hive. This is the work of chalk on slate, the patient turning of a page, the slow accumulation of light in a room that was once dark.

II. The Untaught

But we cannot learn the taste of salt on another's grief, nor the true weight of their memory. You cannot learn the exact moment the heart decides to break, or how to rebuild a home from a whisper. No book can teach you the fear of a future you cannot see, the sudden, impossible joy of a single perfect moment. These are not lessons; they are landscapes. They must be walked, not studied. They are a language without words, a curriculum of the blood, and we must simply live our way into understanding them.

poem about what learning is

The Architecture of Learning

Learning is not the filling of a bucket, but the slow, deliberate work of building.

It begins not with a word, but with a question, a small stone of "why" dropped into the still water of what you thought you knew. The ripples spread and blur the familiar shape of your world, forcing you to see it new.

It is the patient stacking of blocks, each fact a brick, each experience a mortar. Some days the walls rise straight and true; other days they collapse into a heap of splintered certainty and scattered hope. You learn to build with a different material when the old one fails.

Learning is the slow carving of a channel, the way water finds its path through rock, relentless and quiet. It is the sudden click of a lock, the moment a new idea turns the key and a heavy door swings open, revealing a room you never knew was there.

It is not a destination, but a state of being— a constant, beautiful reconstruction, the quiet architecture of a self that is always in the making.

poem about how learning is defined differently

On Definitions

To the child, learning is a question, a constant hunger for the name of things: *What is that? Why is the sky blue?* It is the world as an endless list of curiosities, a giant, beautiful book with no pages torn, no chapters skipped.

To the student, learning is a weapon, a sharpened blade for the long climb. It is the mastery of dates, of formulas, the precise recitation of what has been. It is the weight of a textbook, a frantic hand racing across a blank page.

To the adult, learning is a wound. It is the sudden, sharp understanding of a mistake you've made, the painful truth of a friendship lost, the slow, quiet erosion of a long-held belief. It is the unlearning of everything you thought was true.

To the old soul, learning is a harvest. It is the quiet summation of all that has been— the good, the bad, the beautiful, the broken. It is the slow turning of memories into wisdom, the gentle understanding that the world is not a puzzle to be solved, but a mystery to be held.

Made in United States
Troutdale, OR
11/03/2026

56845114R00106